3 [redacted]02

D1243311

Data Science Revealed

With Feature Engineering, Data Visualization, Pipeline Development, and Hyperparameter Tuning

COMPUTERS

004
NOK

Nokeri, Tshepo Chris

Data science revealed

5/18/2021

Tshepo Chris Nokeri

Apress®

Data Science Revealed

Tshepo Chris Nokeri
Pretoria, South Africa

ISBN-13 (pbk): 978-1-4842-6869-8 ISBN-13 (electronic): 978-1-4842-6870-4
https://doi.org/10.1007/978-1-4842-6870-4

Copyright © 2021 by Tshepo Chris Nokeri

This work is subject to copyright. All rights are reserved by the Publisher, whether the whole or part of the material is concerned, specifically the rights of translation, reprinting, reuse of illustrations, recitation, broadcasting, reproduction on microfilms or in any other physical way, and transmission or information storage and retrieval, electronic adaptation, computer software, or by similar or dissimilar methodology now known or hereafter developed.

Trademarked names, logos, and images may appear in this book. Rather than use a trademark symbol with every occurrence of a trademarked name, logo, or image we use the names, logos, and images only in an editorial fashion and to the benefit of the trademark owner, with no intention of infringement of the trademark.

The use in this publication of trade names, trademarks, service marks, and similar terms, even if they are not identified as such, is not to be taken as an expression of opinion as to whether or not they are subject to proprietary rights.

While the advice and information in this book are believed to be true and accurate at the date of publication, neither the authors nor the editors nor the publisher can accept any legal responsibility for any errors or omissions that may be made. The publisher makes no warranty, express or implied, with respect to the material contained herein.

Managing Director, Apress Media LLC: Welmoed Spahr
Acquisitions Editor: Celestin Suresh John
Development Editor: Laura Berendson
Coordinating Editor: Aditee Mirashi

Cover designed by eStudioCalamar

Cover image designed by Freepik (www.freepik.com)

Distributed to the book trade worldwide by Springer Science+Business Media New York, 1 New York Plaza, Suite 4600, New York, NY 10004-1562, USA. Phone 1-800-SPRINGER, fax (201) 348-4505, e-mail orders-ny@springer-sbm.com, or visit www.springeronline.com. Apress Media, LLC is a California LLC and the sole member (owner) is Springer Science + Business Media Finance Inc (SSBM Finance Inc). SSBM Finance Inc is a **Delaware** corporation.

For information on translations, please e-mail booktranslations@springernature.com; for reprint, paperback, or audio rights, please e-mail bookpermissions@springernature.com.

Apress titles may be purchased in bulk for academic, corporate, or promotional use. eBook versions and licenses are also available for most titles. For more information, reference our Print and eBook Bulk Sales web page at www.apress.com/bulk-sales.

Any source code or other supplementary material referenced by the author in this book is available to readers on GitHub via the book's product page, located at www.apress.com/978-1-4842-6869-8. For more detailed information, please visit www.apress.com/source-code.

Printed on acid-free paper

I dedicate this book to my family and everyone who merrily played influential roles in my life.

Table of Contents

About the Author

Tshepo Chris Nokeri harnesses advanced analytics and artificial intelligence to foster innovation and optimize business performance. In his functional work, he delivered complex solutions to companies in the mining, petroleum, and manufacturing industries. He initially completed a bachelor's degree in information management. Afterward, he graduated with an honor's degree in business science at the University of the Witwatersrand on a TATA Prestigious Scholarship and a Wits Postgraduate Merit Award. They unanimously awarded him the Oxford University Press Prize.

About the Technical Reviewer

Manohar Swamynathan is a data science practitioner and an avid programmer, with more than 14 years of experience in various data science–related areas that include data warehousing, business intelligence (BI), analytical tool development, ad hoc analysis, predictive modeling, data science product development, consulting, and formulating strategy and executing analytics programs. He has had a career covering the life cycle of data across different domains such as US mortgage banking, retail/e-commerce, insurance, and industrial IoT. He has a bachelor's degree with a specialization in physics, mathematics, and computers, and a master's degree in project management. He's currently living in Bengaluru, the Silicon Valley of India.

Acknowledgments

This is my first book, which makes it significant to me. Writing a single-authored book is demanding, but I received firm support and active encouragement from my family and dear friends. Many heartfelt thanks to Professor Chris William Callaghan and Mrs. Renette Krommenhoek from the University of the Witwatersrand. They gallantly helped spark my considerable interest in combating practical and real-world problems using advanced analytics. I would not have completed this book without the valuable help of the dedicated publishing team at Apress, which compromises Aditee Mirashi and Celestin Suresh John. They trusted and ushered me throughout the writing and editing process. Last, humble thanks to all of you reading this; I earnestly hope you find it helpful.

Introduction

Welcome to *Data Science Revealed*. This book is your guide to solving practical and real-world problems using data science procedures. It gives insight into data science techniques, such as data engineering and visualization, statistical modeling, machine learning, and deep learning. It has a rich set of examples on how to select variables, optimize hyperparameters, develop pipelines, and train, test and validate machine and deep learning models. Each chapter contains a set of examples allowing you to understand the concepts, assumptions, and procedures behind each model.

First, it conceals the parametric method or linear model and the means for combating underfitting or overfitting using regularization techniques such as lasso and ridge. Next, it concludes complex regression by presenting time-series smoothening, decomposition, and forecasting. Then, it takes a fresh look at a nonparametric model for binary classification, known as logistic regression, and ensemble methods such as decision tree, support vector machine, and naïve Bayes. Next, it covers the most popular nonparametric method for time-event data, recognized as the Kaplan-Meier estimator. It also covers ways of solving a classification problem using artificial neural networks, like the restricted Boltzmann machine, multilayer perceptron, and deep belief network. Then, it summarizes unsupervised learning by uncovering clustering techniques, such as K-means, agglomerative and DBSCAN, and dimension reduction techniques such as feature importance, principal component analysis, and linear discriminant analysis. In addition, it introduces driverless artificial intelligence using H2O.

It uses Anaconda (an open source distribution of Python programming) to prepare the examples. The following are some of the libraries covered in this book:

- Pandas for data structures and tools
- Statsmodels for basic statistical computation and modeling
- SciKit-Learn for building and validating key machine learning algorithms
- Prophet for time-series analysis
- Keras for high-level frameworks for deep learning

- H2O for driverless machine learning
- Lifelines for survival analysis
- NumPy for arrays and matrices
- SciPy for integrals, solving differential equations and optimization
- Matplotlib and Seaborn for popular plots and graphs

This book targets beginner to intermediate data scientists and machine learning engineers who want to learn the full data science process. Before exploring the contents of this book, ensure that you understand the basics of statistics, Python programming, and probability theories. Also, you'll need to install the packages mentioned in the previous list in your environment.

CHAPTER 1

An Introduction to Simple Linear Regression

This book introduces you to the world of data science. It reveals the proper way to do data science. It covers essential statistical and programming techniques to help you understand data science from a broad perspective. Not only that, but it provides a theoretical, technical, and mathematical foundation for problem-solving using data science techniques.

This chapter covers the *parametric method*, also called the *linear method*. Understanding how to test a regressor under the violation of regression assumptions will enable you to tackle problems in subsequent chapters with ease. While reading, it is important to remember that the example data has one dependent variable. This chapter does not cover multicollinearity with a variance inflation factor (VIF).

Simple linear regression estimates the nature of the relationship between an independent variable and a dependent variable, where an *independent variable* is a continuous variable or a categorical variable and a *dependent variable* is inevitably a continuous variable. It investigates how a change in an independent variable influences a change in a dependent variable. We express the standard formula as shown in Equation 1-1.

$$y = a + bx \qquad \text{(Equation 1-1)}$$

Here, y represents a dependent variable, a represents an intercept (the mean value of a dependent variable given that we hold an independent variable constant), b represents a slope or gradient (the direction of a straight line), and x represents an independent variable.

The model fits a straight line to the data points. If it perfectly fits a straight line to the data points, then it is an exemplary model. In contrast, if the data points severely deviate away from a straight line, then it violates regression assumptions.

© Tshepo Chris Nokeri 2021
T. C. Nokeri, *Data Science Revealed*, https://doi.org/10.1007/978-1-4842-6870-4_1

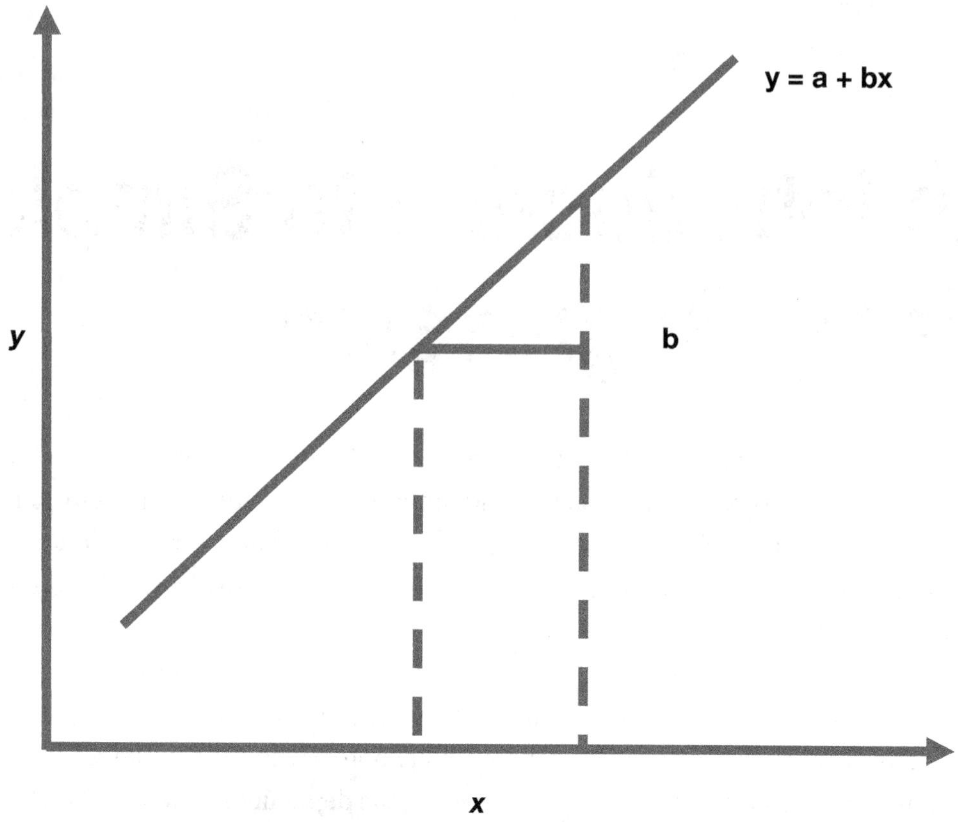

***Figure 1-1.** Pairwise scatter plot for a perfect linear regression modelx*

Figure 1-1 illustrates an optimal model (it hardly happens in the actual world because of variability). At most, the data points scatter close to the straight line. To combat this problem, a regressor (or a regression model) introduces an error term during modeling so that the squared deviation of the data points is small. It estimates an intercept and a slope to reduce the error terms. The most common regressor is the least-squares model. Equation 1-2 expresses the formula of the least-squares model.

$$\hat{y} = \beta_0 + \beta_1 X_1 + \varepsilon_i \qquad \text{(Equation 1-2)}$$

Here, $\hat{y}$ represents an expected dependent variable, β_0 represents an intercept, β_1 represents a slope, X_1 represents an independent variable, and ε_i represents the error terms (the residual for the ith of n data points) expressed as shown in Equation 1-3.

$$e_i = y_i - \widehat{y_i} \qquad \text{(Equation 1-3)}$$

The least-squares regressor ensures that the sum squares of residuals are small. Equation 1-4 estimates the sum squares of residuals by using the property underneath.

$$e_1^2 + e_2^2 \ldots e_i^2 \qquad \text{(Equation 1-4)}$$

Equation 1-4 assumes that residuals are always equal to zero and that estimates are unbiased.

Regression Assumptions

If a regressor satisfies the assumptions specified in Table 1-1, then it is reliable.

Table 1-1. *Linear Regression Assumptions*

Assumption	Description
Linearity	There must be a linear relationship between an independent variable and a dependent variable. We verify this assumption using a pairwise scatter plot.
Normality	Data must come from a normal distribution. There is a normal distribution when values of a variable spread symmetrically around the true mean value. We verify this assumption using a normal probability plot (also called a *normal Q-Q plot*). For a reliable test of normality, use a test statistic called the *Kolmogorov-Smirnov test*. If there is no normality, then transform the data.
Multicollinearity	There must be little or no multicollinearity in the data. If an independent variable has a high correlation, then there is multicollinearity. We verify this assumption by using a correlation matrix, tolerance, and VIF. If there is multicollinearity, we must center the data. The example data has one dependent variable, so multicolllinearity with VIF is not covered.
No autocorrelation	There must be no autocorrelation in the residuals. There is autocorrelation when residuals are not independent of each other. We verify this assumption using a lag plot, autocorrelation function plot, and partial autocorrelation plot. For a reliable test of autocorrelation, use the Dublin-Watson test.

In contrast, if a regressor violates the underlying regression assumption, then it becomes difficult to generalize a model.

Simple Linear Regression in Practice

The example data[1] has two columns (each column represents a variable). The first column contains the values of employees' years of work experience (an independent variable), and the second column contains the values of employees' salaries (a dependent variable).

Check Data Quality

The least-squares model is sensitive (the quality of the data influences its performance). For optimal model performance, ensure there are no missing values or outliers and scale the data prior to training the regressor.

Detect Missing Values

The visible presence of missing values in the data results in poor model performance. To avoid this dilemma, after loading the data, check whether there are missing values in the data.

Missing Values Heatmap

Listing 1-1 plots a heatmap that shows the missing values (see Figure 1-2).

Listing 1-1. Missing Values Heatmap

```
import seaborn as sns
sns.heatmap(df.isnull(),cmap="Blues")
plt.show()
```

[1] https://www.kaggle.com/rohankayan/years-of-experience-and-salary-dataset

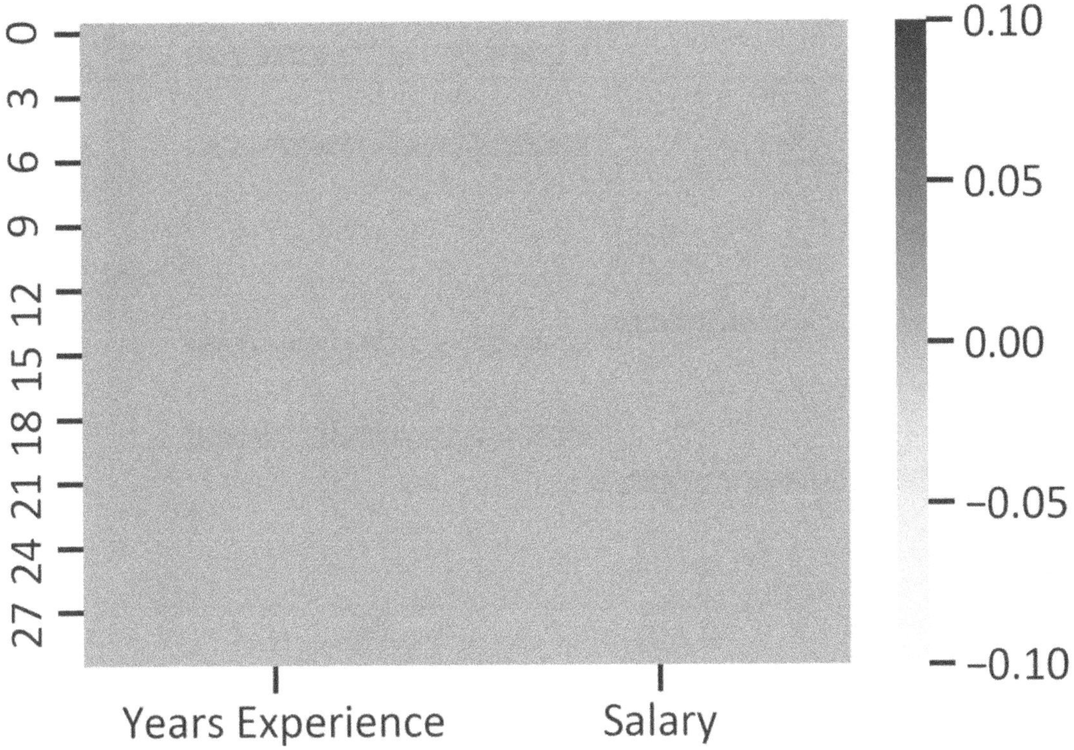

Figure 1-2. *Missing values heatmap*

Figure 1-2 shows that there are no missing values detected in the data. If there are missing values in the data, the heatmap looks like Figure 1-3.

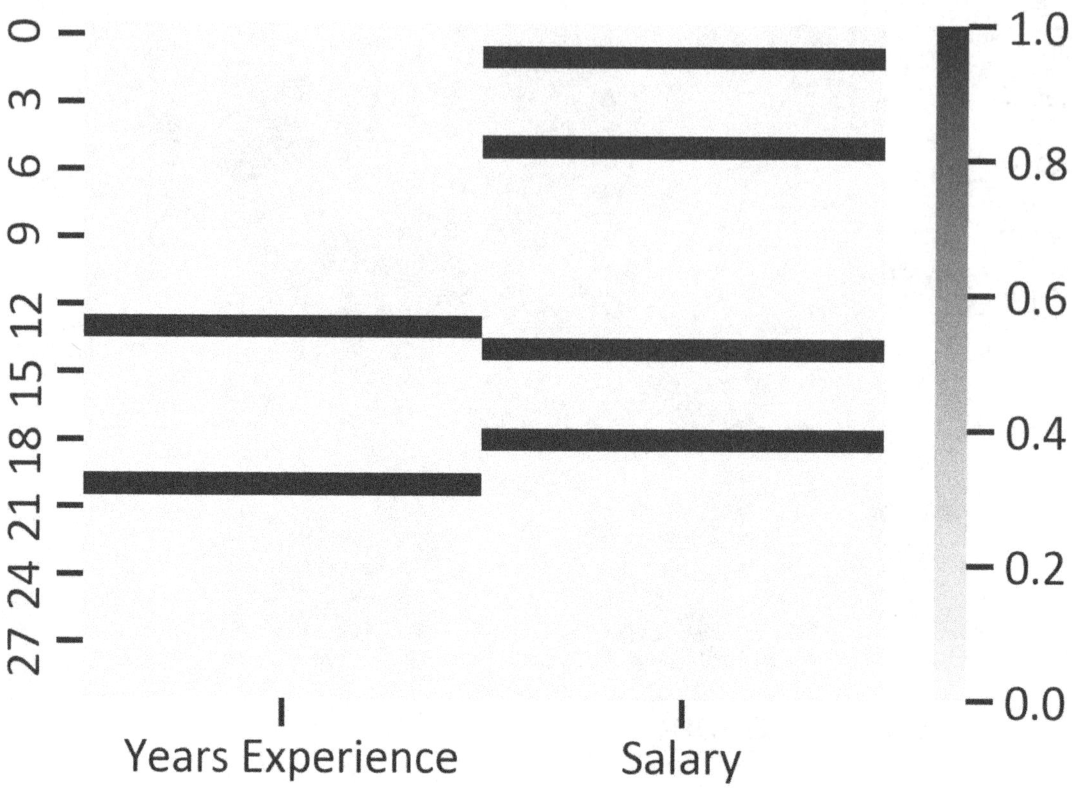

Figure 1-3. *Missing values heatmap*

If there are missing values in the data, replace the missing values with the mean value or the median of the variable. Listing 1-2 replaces the missing values with the mean value.

Listing 1-2. Replace Missing Values with the Mean Value

```
df["YearsExperience"].fillna(df["YearsExperience"].mean())
df["Salary"].fillna(df["YearsExperience"].mean())
```

Listing 1-3 replaces missing values with the median value.

Listing 1-3. Replace Missing Values with the Median Value

```
df["YearsExperience"].fillna(df["YearsExperience"].median())
df["Salary"].fillna(df["YearsExperience"].median())
```

Detect Normality

Regressors assume the data follows a normal distribution (data points are spread symmetrically around the true mean value). If the data is not normal, perform data transformation to reduce its skewness. Negatively skewed data requires a power transformation or an exponential transformation. In contrast, positively skewed data requires a log transformation or square root transformation. A standardized normal distribution with different means and variances is written mathematically in terms of the distribution.

$$N(0,1) \qquad \text{(Equation 1-5)}$$

Here, the mean value is equal to 0, and the standard deviation is equal to 1.

Histogram

A histogram plot has the intervals of a variable on the x-axis and the frequency of the values on the y-axis. Listing 1-4 plots the employee's years of work experience, and Figure 1-4 exhibits the distribution of employees' years of work experience.

Listing 1-4. Employees' Years of Work Experience Histogram

```
sns.distplot(df["YearsExperience"])
plt.ylabel("Frequency")
plt.xlabel("Experience")
plt.show()
```

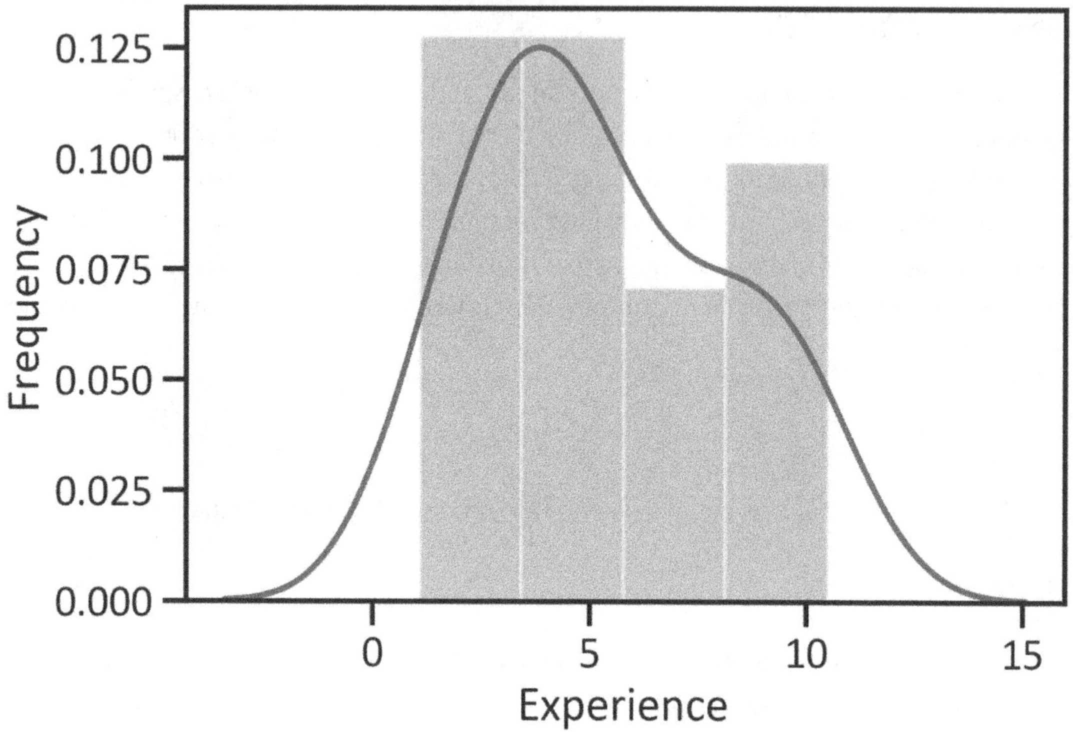

Figure 1-4. *Histogram*

Figure 1-4 shows that the values of employees' years of work experience slightly follow a normal distribution. We cannot summarize the independent variable using the mean value.

Listing 1-5 plots the histogram for employees' salaries (see Figure 1-5).

Listing 1-5. Employees' Salaries Histogram

```
sns.distplot(df["Salary"])
plt.ylabel("Frequency")
plt.xlabel("Salary")
plt.show()
```

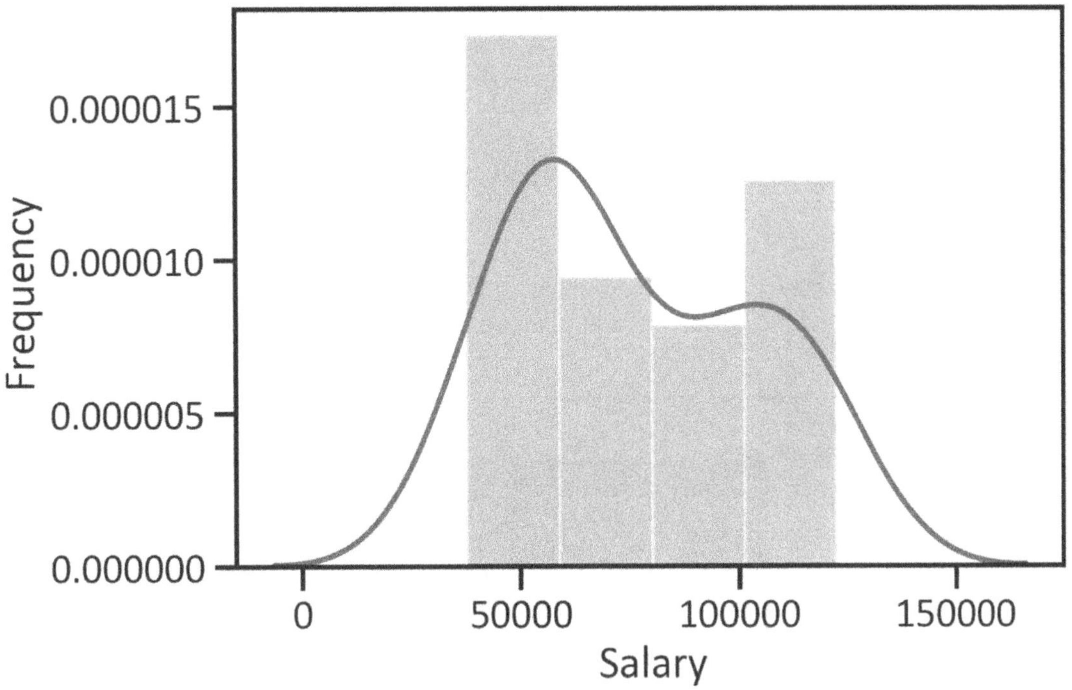

Figure 1-5. *Histogram*

Both Figure 1-4 and Figure 1-15 show nonextreme normal distributions. Skewness frequently occurs when there are outliers in the data.

Detect Outliers

An outlier represents a data point that is too small or large. It can influence the model by inflating error rates. If there are outliers in the data, remove them, or replace them with the mean value or median value.

Box Plot

The simplest way to detect outliers is by using a box plot. A box plot succinctly summarizes information about the distinctive shape and dispersion of the data. It has the following properties: a box representing the middle values in the data, the median line representing the point where 50% of the data points are, and quartiles representing the point at which 25% of the data points are above and 75% of the data points are below the straight line. While Q3 represents the point at which 75% of the data is over and 25%

of it is under, the straight line and the other lines are representing a whisker that joins Q1 or Q3 with the farthest data point other than an outlier.

Listing 1-6 plots the employees' years of work experience.

Listing 1-6. Employees' Years of Work Experience Box Plot

```
sns.boxplot(df["YearsExperience"])
plt.xlabel("Experience")
plt.show()
```

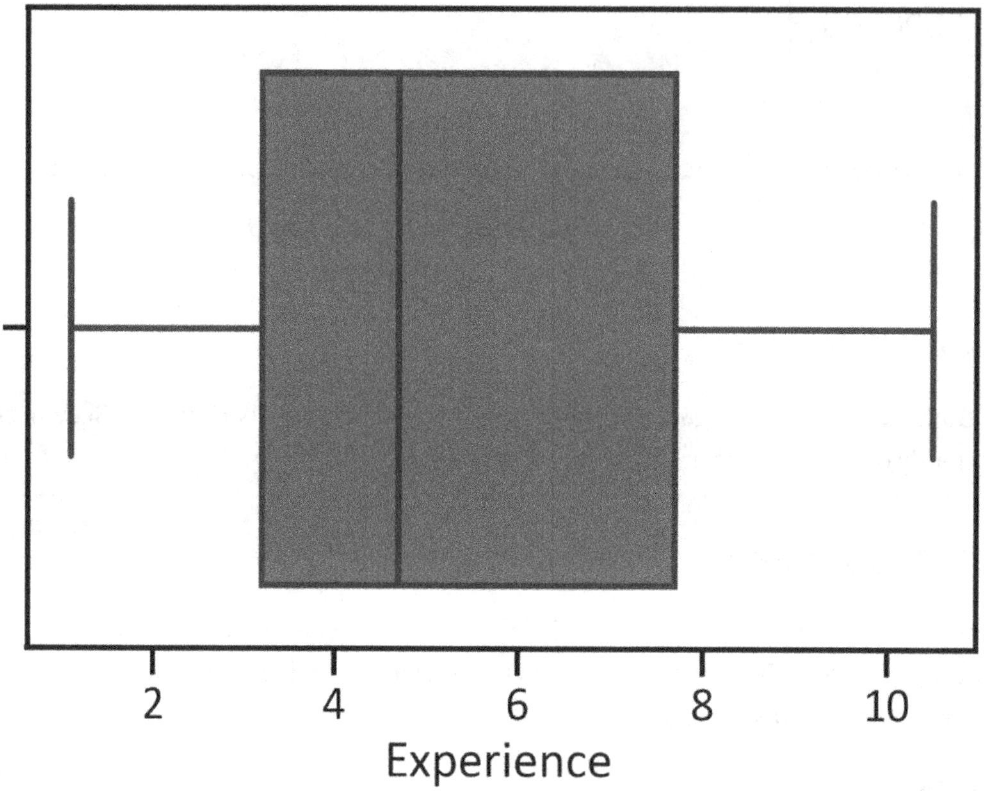

Figure 1-6. *Box plot*

Figure 1-6 shows that the values of employees' years of work experience are slightly skewed to the left, and there were no outliers in the data.

Listing 1-7 constructs a box plot for the values of the dependent variable (see Figure 1-7).

Listing 1-7. Employees' Salaries Box Plot

```
sns.boxplot(df["Salary"])
plt.xlabel("Salary")
plt.show()
```

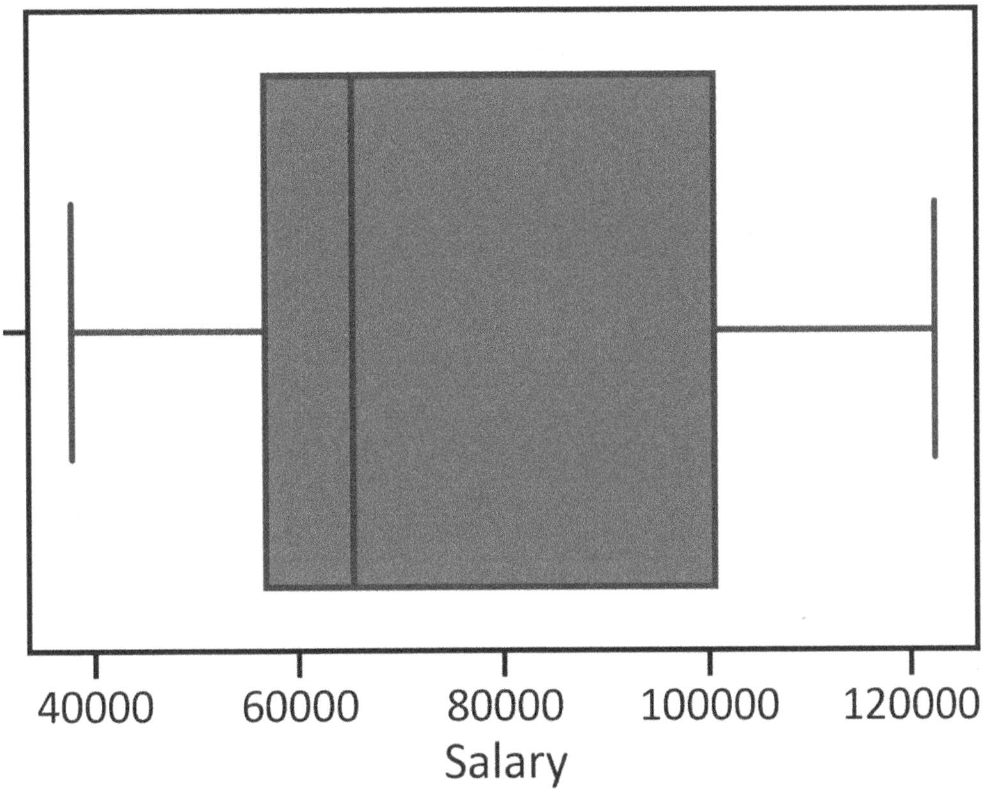

Figure 1-7. *Box plot*

Figure 1-7 indicates that the values of employees' salaries are skewed to the left, and there are no outliers detected in the data. Both Figure 1-6 and Figure 1-7 slightly resemble a normal distribution.

Listing 1-8 plots the employees' years of work experience and employees' salaries. Figure 1-8 exhibits changes over time connected by a straight line. It has two axes: x-axis (horizontal) and y-axis (vertical). The axes are graphically denoted as (x, y). The independent variable is on the x-axis, and the dependent is on the y-axis.

Listing 1-8. Employees' Years of Work Experience and Employees' Salaries Line Plot

```
df.plot(kind="line")
plt.xlabel("Experience")
plt.ylabel("Salary")
plt.show()
```

Figure 1-8. *Line plot*

Figure 1-8 does not show an ideal line. The data points invariably follow a long-run upward trend. The straight line slopes upward and to the right. Employees' salaries increase as their years of work experience increase.

Listing 1-9 returns a plot that shows the probability density function using kernel density estimation (see Figure 1-9). It captures the probabilities of the continuous random variable.

Listing 1-9. Density Plot

```
sns.kdeplot(df["YearsExperience"],df["Salary"])
plt.title("Experience and salary density plot")
plt.xlabel("Experience")
plt.ylabel("Salary")
plt.show()
```

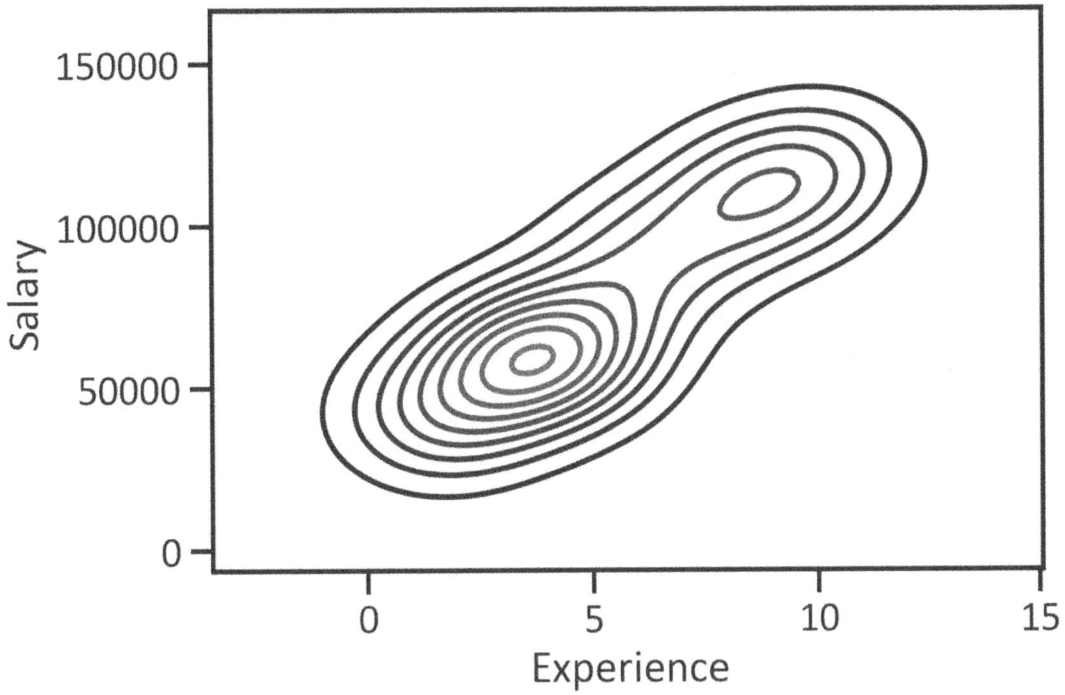

Figure 1-9. *Density plot*

Figure 1-9 suggests a two-dimensional probability density; however, the distribution is not normal enough. It has clustered data points, making it inadvertently difficult to interpret the data.

Listing 1-10 returns a scatter plot that depicts the association between variables (see Figure 1-10).

Listing 1-10. Employees' Years of Work Experience and Employees' Salaries Joint Plot

```
sns.jointplot(x="YearsExperience",y="Salary",data=df,kind="reg")
plt.xlabel("Experience")
plt.ylabel("Salary")
plt.show()
```

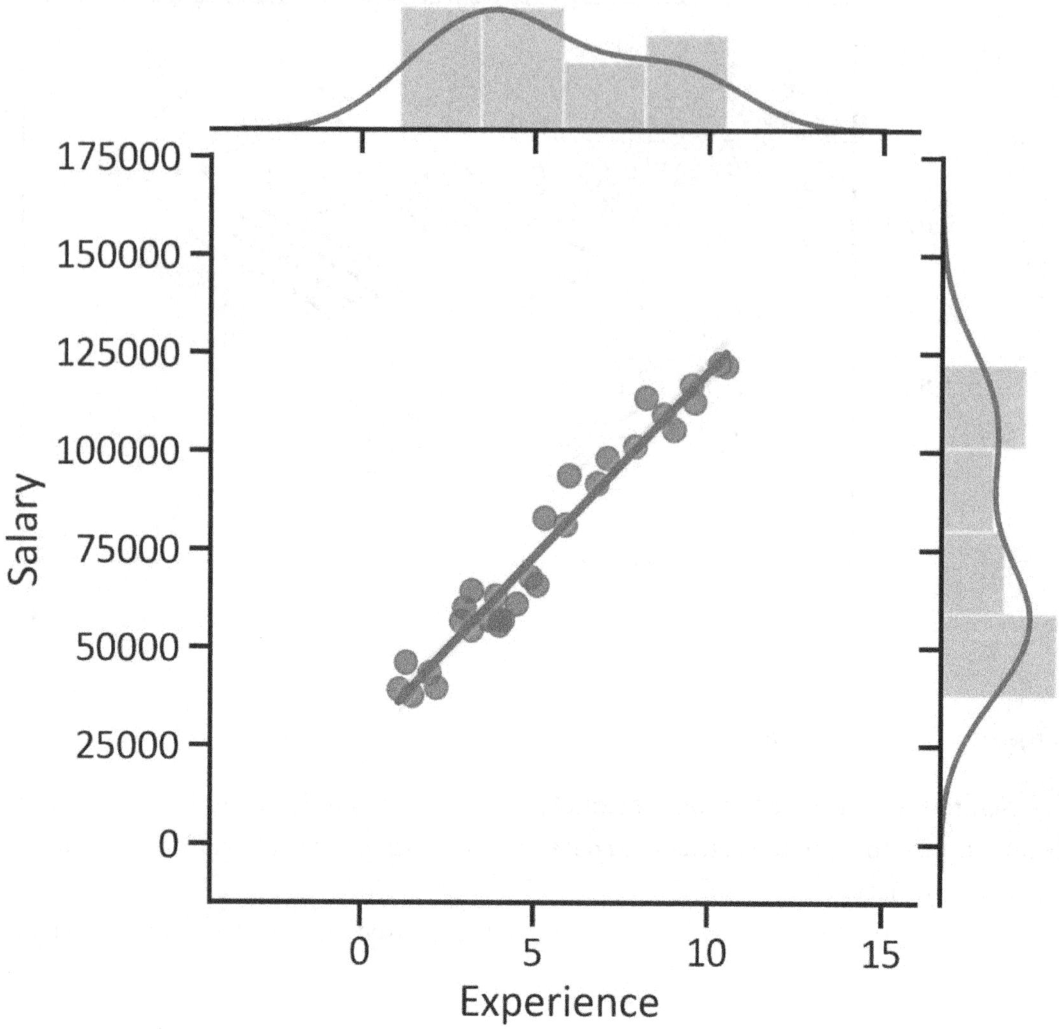

Figure 1-10. *Pairwise scatter plot*

Figure 1-10 indicates that a straight line perfectly fits the data points. There is a strong positive correlation relationship between employees' years of work experience and employees' salaries.

Listing 1-11 summarizes the data using the central tendency (see Table 1-2).

Listing 1-11. Summarize the Central Tendency

```
df.describe().transpose()
```

Table 1-2. *Descriptive Statistics*

	count	mean	std	min	25%	50%	75%	max
YearsExperience	30.0	5.313333	2.837888	1.1	3.20	4.7	7.70	10.5
Salary	30.0	76003.000000	27414.429785	37731.0	56720.75	65237.0	100544.75	122391.0

Table 1-2 highlights that the arithmetic average of employees' work experience is 5 years and that of employees' salaries is $76,003 per year. The values of employees' years of work experience deviate away from the mean value by 3, and the values of employees' salaries deviate by $27,414. The lowest number of work year experience is 1 year, and the lowest salary is $37,731. An employee with the highest salary earns $122,391 annually and has about 11 years of work experience. One out of four employees earns a salary of about $56,721 annually. Employees who earn a salary of $100,544 and above annually fall under the upper echelon.

Understand Correlation

Linear regression is all about examining correlation coefficients. Correlation estimates the apparent strength of a linear relationship between an independent variable and a dependent variable. There are several methods for determining the correlation between two variables like the Pearson correlation method, Spearman correlation method, and Kendall correlation method.

The Pearson Correlation Method

Listing 1-12 applies the Pearson correlation method (the dependent variable is a continuous variable). Equation 1-6 expresses the method.

$$r_{xy} = \frac{s_{xy}}{s_x s_y} \qquad \text{(Equation 1-6)}$$

Here, s_x represents the standard deviation of an independent variable, s_y represents the standard deviation of a dependent variable, and s_{xy} represents the covariance. The method produces values that range from -1 to 1, where -1 shows a strong negative correlation relationship, 0 shows no correlation relationship, and 1 shows a strong positive correlation relationship.

Correlation Matrix

Figure 1-11 shows the correlation between employees' years of work experience and employees' salaries.

Listing 1-12. Correlation Matrix

```
dfcorr = df.corr(method="pearson")
sns.heatmap(dfcorr, annot=True,annot_kws={"size":12}, cmap="Blues")
plt.show()
```

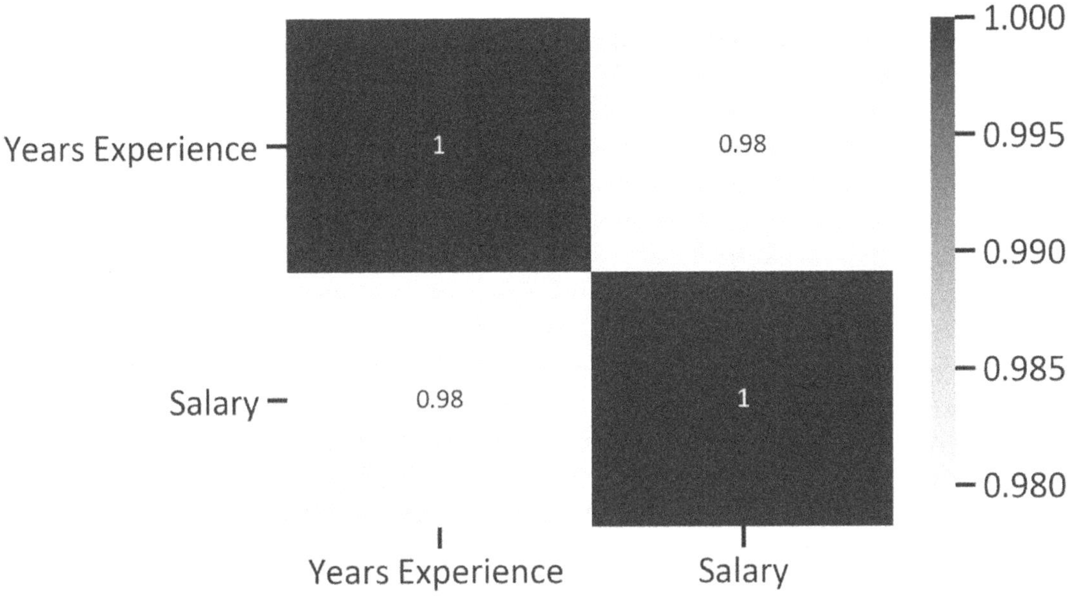

Figure 1-11. *Correlation matrix heatmap*

Figure 1-11 shows a line of 1s that go from the top left to the bottom right (each variable perfectly correlates with itself). There is strong positive correlation between the employees' years of work experience and the employees' salaries.

The Covariance Method

Covariance is the joint variability between the two variables (it estimates how two variables vary together). Equation 1-7 expresses the covariance method.

$$S_{xy} = \frac{\Sigma(x_i - \underline{x})(y_i - \underline{y})}{n-1} \quad \text{(Equation 1-7)}$$

Here, n is the number of the samples, y_i represents scalar random variables, and $\underline{x}$ and $\underline{y}$ are the mean values of the scalar random variables.

Covariance Matrix

Listing 1-13 produces the covariance matrix (see Figure 1-12).

Listing 1-13. Covariance Matrix

```
dfcov = df.cov()
sns.heatmap(dfcov, annot=True,annot_kws={"size":12}, cmap="Blues")
plt.show()
```

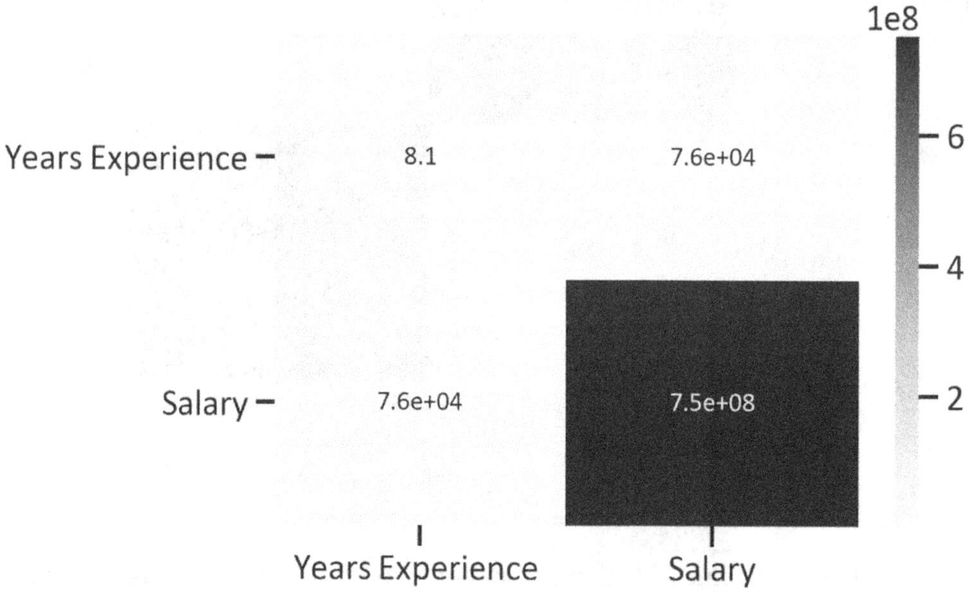

Figure 1-12. *Covariance matrix heatmap*

Figure 1-12 reveals that the covariance is approximately 0.0076 (which is close to 1). It confirms a positive relationship between variables.

Assign and Reshape Arrays

Listing 1-14 assigns arrays of *x* and *y* using NumPy. An array holds one or more variables.

Listing 1-14. Assign x and y arrays

```
x = np.array(df["YearsExperience"])
y = np.array(df["Salary"])
```

Listing 1-15 reshapes *x* and *y* (transforms one-dimensional data into two-dimensional data).

Listing 1-15. Reshape x and y arrays

```
x = x.reshape(-1,1)
y = y.reshape(-1,1)
```

Split Data into Training and Test Data

Supervised models require a data partition. Listing 1-16 splits the data into training and test data by calling the `train_test_split()` method.

Listing 1-16. Split Data into Training and Test Data

```
from sklearn.model_selection import train_test_split
x_train, x_test, y_train, y_test = train_test_split(x,y,test_size=0.2,
random_state=0)
```

Note that 80% of the data is for training, and 20% of the data is for testing.

Normalize Data

Widespread scaling methods include `StandardScaler()`, which scales data in such a way that the mean value is 0 and the standard deviation is 1, 2); `MinMaxScaler()`, which scales data between 0 and 1 or between -1 and 1 if there are negative values in the data; and `RobustScaler()`, which removes outliers and uses scale data (using either of the previous scales). Listing 1-17 applies the `StandardScaler()` method to normalize data.

Listing 1-17. Normalize Data

```
from sklearn.model_selection import StandardScaler
scaler = StandardScaler()
x_train = scaler.fit_transform(x_train)
x_test = scaler.transform(x_test)
```

Develop the Least Squares Model Using Statsmodels

Statsmodels is a Python scientific package that contains a set of distinct classes and functions for correctly estimating statistical models and performing test statistics. It complements the SciPy package for statistical computation. By default, it does not include an intercept. Listing 1-18 manually adds an intercept to the model.

Listing 1-18. Add a Constant

```
x_constant = sm.add_constant(x_train)
x_test = sm.add_constant(x_test)
```

Listing 1-19 completes the regressor.

Listing 1-19. Develop the Least Squares Model

```
model = sm.OLS(y_train, x_constant).fit()
```

Predictions

Listing 1-20 tabulates the predicted values (see Table 1-3).

Listing 1-20. Predicted Values

```
y_pred = model.predict(x_test)
pd.DataFrame(y_pred, columns = ["Predicted salary"])
```

Table 1-3. *Predicted Values*

	Predicted Salary
0	40748.961841
1	122699.622956
2	64961.657170
3	63099.142145
4	115249.562855
5	107799.502753

Evaluate the Model

Listing 1-21 returns a summary of the performance of the regressor (see Table 1-4).

Listing 1-21. Profile

```
summary = model.summary()
summary
```

Table 1-4. *Results*

Dep. Variable:	y	**R-squared:**	0.941
Model:	OLS	**Adj. R-squared:**	0.939
Method:	Least Squares	**F-statistic:**	352.1
Date:	Fri, 16 Oct 2020	**Prob (F-statistic):**	5.03e-15
Time:	14:42:15	**Log-Likelihood:**	-242.89
No. Observations:	24	**AIC:**	489.8
Df Residuals:	22	**BIC:**	492.1
Df Model:	1		
Covariance Type:	nonrobust		

	coef	**std err**	**T**	**P>|t|**	**[0.025**	**0.975]**
const	7.389e+04	1281.861	57.640	0.000	7.12e+04	7.65e+04
x1	2.405e+04	1281.861	18.765	0.000	2.14e+04	2.67e+04

Omnibus:	3.105	**Durbin-Watson:**	2.608
Prob(Omnibus):	0.212	**Jarque-Bera (JB):**	1.567
Skew:	0.297	**Prob(JB):**	0.457
Kurtosis:	1.898	**Cond. No.**	1.00

R-squared represents the variation the model explains about the data. On the other hand, adjusted R-squared represents the variation of the independent variable that influences the dependent variable explains. The model explains 94.1% of the variation in the data, and the independent variable (YearsExperience) explains 93.9% of the variation in the data. The model best explains the data. The p-value was barely 0.05. We reject the null hypothesis in favor of the alternative hypothesis. There is a significant difference between the employees' years of work experience and the employees' salaries.

Develop the Least Squares Model Using SciKit-Learn

The SciKit-Learn package is popular for training and testing models. It has features that enable data preprocessing and model selection, training, and evaluation.

Least Squares Model Hyperparameter Optimization

A hyperparameter is a value set before training a model. It controls a model's learning process. Hyperparameter optimization involves specifying a list of values and finding values that yield optimal model performance.

Step 1: Fit the Least Squares Model with Default Hyperparameters

Listing 1-22 fits the least squares model with default hyperparameters.

Listing 1-22. Develo the Least Squares Model

```
from sklearn.linear_model import LinearRegression
lm = LinearRegression()
lm.fit(x_train,y_train)
```

Step 2: Determine the Mean and Standard Deviation of the Cross-Validation Scores

Cross-validation involves examining the extent to which the model generalizes the data. Listing 1-24 applies the R^2 score as a criterion for finding the cross-validation scores. There are other scores that one can use as a criterion such as the mean squared error (the variability explained by the model about the data after considering a regression relationship) and the root mean squared error (the variability explained without considering a regression relationship). Listing 1-23 finds the default parameters of the regressor.

Listing 1-23. Default Parameters

```
lm.get_params()
```

An intensive search returns an estimator, a parameter space, the method used to search candidates, the cross-validation scheme, and the score function. Cross-validation methods include RandomizedSearchCV, which considers a specific number of candidates; GridSearchCV, which considers all parameter combinations; and BayesSearchCV, which uses previous loss to determine the best point to sample the loss by processing the Gaussian process as a prior for function. Underneath, we show you how to perform cross-validation using GridSearchCV and BayesSearchCV. Table 1-5 highlights optimizable hyperparameter.

Table 1-5. *Optimizable Hyperparameters*

Parameter	Description
fit_intercept	Determines whether the model must estimate an intercept
normalize	Determines whether we must normalize the independent variables
copy_X	Determines whether we must copy the independent variable

Listing 1-24 creates the grid model.

Listing 1-24. Develop the Grid Model Using GridSearchCV

```
from sklearn.model_selection import GridSearchCV
param_grid = {'fit_intercept':[True,False],
              'normalize':[True,False],
              'copy_X':[True, False]}
grid_model  = GridSearchCV(estimator=lm,
                           param_grid=param_grid,
                           n_jobs=-1)
grid_model.fit(x_train,y_train)
```

Listing 1-25 returns the best cross-validation and best parameters.

Listing 1-25. Find Best Hyperparameters Using GridSearhCV

```
print("Best score: ", grid_model.best_score_, "Best parameters: ",
grid_model.best_params_)
```

Best score: 0.9272138118711238 Best parameters: {'copy_X': True, 'fit_intercept': True, 'normalize': True}

Another way of finding the right hyperparameters involves using Bayesian optimization. Listing 1-26 performs Bayesian optimization using Skopt. To install it in the Python environment, use `pip install scikit-optimize`.

Listing 1-26. Develop the Grid Model Using BayesSearchCV

```
from skopt import BayesSearchCV
param_grid = {'fit_intercept':[True,False],
              'normalize':[True,False],
              'copy_X':[True, False]}
grid_model = BayesSearchCV(lm,param_grid,n_iter=30,random_state=1234,
verbose=0)
grid_model.fit(x_train,y_train)
```

Listing 1-27 finds the best cross-validation score and best parameters.

Listing 1-27. Find Best Hyperparameters Using BayesSearchCV

```
print("Best score: ", grid_model.best_score_, "Best parameters: ",
grid_model.best_params_)
```

Best score: 0.9266258295344271 Best parameters: OrderedDict([('copy_X', False), ('fit_intercept', True), ('normalize', False)])

Finalize the Least Squares Model

Listing 1-28 completes the least squares model using the hyperparameters estimated in Listing 1-27.

Listing 1-28. Finalize the Least Squares Model

```
lm = LinearRegression(copy_X= True,
                      fit_intercept= True,
                      normalize= True)
lm.fit(x_train,y_train)
```

Find the Intercept

Listing 1-29 estimates the intercept. An intercept is the mean value of an independent variable, given that we hold a dependent variable constant.

Listing 1-29. Intercept

```
lm.intercept_
   array([73886.20833333])
```

Find the Estimated Coefficient

Listing 1-30 estimates coefficients.

Listing 1-30. Coefficient

```
lm.coef_
array([[24053.85556857]])
```

A formula of a straight line is expressed as follows:

$$\hat{y} = 73\,886.21 + 24\,053.851X_1 + \varepsilon_i \qquad \text{(Equation 1-8)}$$

For every additional year of work experience, the salary increases by 24,054.

Test the Least Squares Model Performance Using SciKit-Learn

Listing 1-31 applies the `predict()` method to return predicted values, and then it passes the array to a dataframe (see Table 1-6).

Listing 1-31. Tabulate Predicted Values

```
y_pred = lm.predict(x_test)
pd.DataFrame(y_pred, columns = ["Predicted salary"])
```

Table 1-6. *Actual Values*

	Predicted Salary
0	40748.961841
1	122699.622956
2	64961.657170
3	63099.142145
4	115249.562855
5	107799.502753

Listing 1-32 tabulates actual values (see Table 1-7).

Listing 1-32. Develop an Actual Values Table

```
pd.DataFrame(y_test, columns = ["Actual salary"])
```

Table 1-7. *Actual Values of Salary*

	Actual Salary
0	37731.0
1	122391.0
2	57081.0
3	63218.0
4	116969.0
5	109431.0

Table 1-7 suggests that the regressor makes slight errors.

Mean Absolute Error

An absolute error is a difference between the estimated value and the true value. For example, if a model predicts that the salary is $40,749 while the actual value is $37,731, then the model has an absolute error of $40,749 - $37,731 = $3,018. Mean absolute error (MAE) represents the average magnitude of error in estimates without considering the direction.

$$MAE = \frac{1}{n}\sum_{i=1}^{n}|x_i - x| \quad \text{(Equation 1-9)}$$

Here, n represents the number of errors, and $|x_i - x|$ represents the absolute errors.

Mean Squared Error

Mean squared error (MSE) represents the variability explained by the model about the data after considering a regression relationship.

$$MSE = \frac{1}{n}\sum_{i=1}^{n}|y_i - \hat{y}| \quad \text{(Equation 1-10)}$$

Root Mean Squared Error

Root mean squared error (RMSE) represents the variability explained without considering a regression relationship. To get an RMSE score, use the square root of the MSE.

$$RMSE = \sqrt{MSE} \quad \text{(Equation 1-11)}$$

R-squared

R-squared (R^2) represents the variability explained by the model about the data. The metric comprises values that range from 0 to 1, where 0 shows that the regression model poorly explains the variability in the data, and 1 shows that the regression model best explains the data.

$$R^2 = 1 - \frac{\Sigma(y_i - \hat{y})^2}{\Sigma(y_i - \underline{y})^2} \quad \text{(Equation 1-12)}$$

Here, $\hat{y}$ represents the predicted value of y, and $\underline{y}$ represents the mean value of y. The R^2 score must be multiplied by a hundred. For example, if the score is 0.9, then the model explains 90% of the variability in the data.

There are other metrics for model evaluation such as the following:

Explained variance score: A metric that captures a discrepancy between a model and the actual data. If the score is close to 1, then there is a strong correlation.

Mean gamma deviance: A metric that captures the scaling of a dependent variable and estimates the relative error. It is a Tweedie deviance with a power parameter of 2.

Mean Poisson deviance: A metric that captures the scaling of a dependent variable and estimates the relative error. It is a Tweedie deviance with a power parameter of 1.

Listing 1-33 tabulates the evaluation metrics (see Table 1-8).

Listing 1-33. Evaluation Metrics

```
MAE = metrics.mean_absolute_error(y_test,y_pred)
MSE = metrics.mean_squared_error(y_test,y_pred)
RMSE = np.sqrt(MSE)
R2 = metrics.r2_score(y_test,y_pred)
EV = metrics.explained_variance_score(y_test,y_pred)
MGD = metrics.mean_gamma_deviance(y_test,y_pred)
MPD = metrics.mean_poisson_deviance(y_test,y_pred)
lmmodelevaluation = [[MAE,MSE,RMSE,R2,EV,MGD,MPD]]
lmmodelevaluationdata = pd.DataFrame(lmmodelevaluation,
                                     index = ["Values"],
                                     columns = ["MAE",
                                              "MSE",
                                              "RMSE",
                                              "R2",
                                              "Explained variance score",
                                              "Mean gamma deviance",
                              "Mean Poisson deviance"]).transpose()
lmmodelevaluationdata
```

Table 1-8. Performance Matrix

	Values
MAE	2.446172e+03
MSE	1.282341e+07
RMSE	3.580979e+03
R^2	9.881695e-01
Explained variance score	9.897038e-01
Mean gamma deviance	3.709334e-03
Mean Poisson deviance	2.129260e+02

Table 1-8 indicates that the model explains 98% of the variability of the data. There is a strong correlation between employees' years of work experience and employees' salaries. On average, the magnitude of errors without considering the direction is 2,446, and the mean squared errors is 1.

Training Data

Listing 1-34 verifies whether there is a linear relationship between the variables by plotting a line through the data points (see Figure 1-13).

Listing 1-34. Training Data

```
plt.scatter(x_train,y_train,s=200)
plt.plot(x_test,y_pred,color="red")
plt.xlabel("Actual Experience")
plt.ylabel("Actual Salary")
plt.show()
```

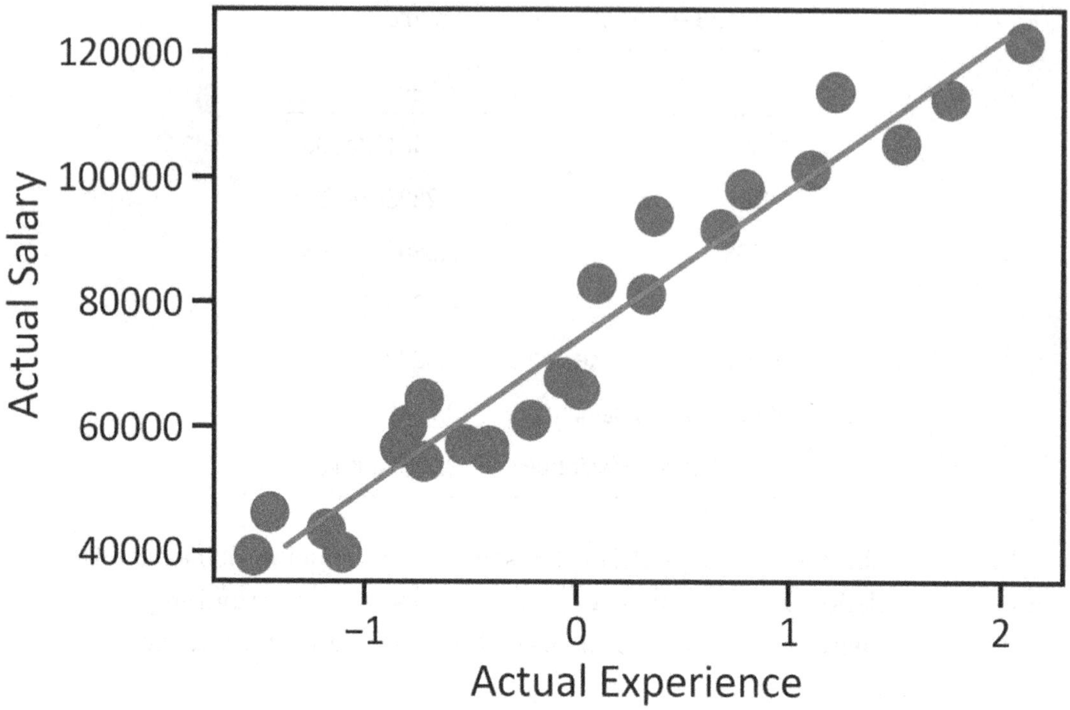

Figure 1-13. *Training data*

Figure 1-13 confirms that the data points are close to a straight line, but the closeness is not extreme. There is a linear relationship between employees' years of work experience and employees' salaries.

Test Data

Listing 1-35 returns Figure 1-14, which shows the actual values of employees' years of work experience and predicted values of employees' years of work experience.

Listing 1-35. Test Data

```
plt.scatter(y_test,y_pred,s=200)
plt.axhline(color="red")
plt.xlabel("Actual Salary")
plt.ylabel("Predicted Salary")
plt.show()
```

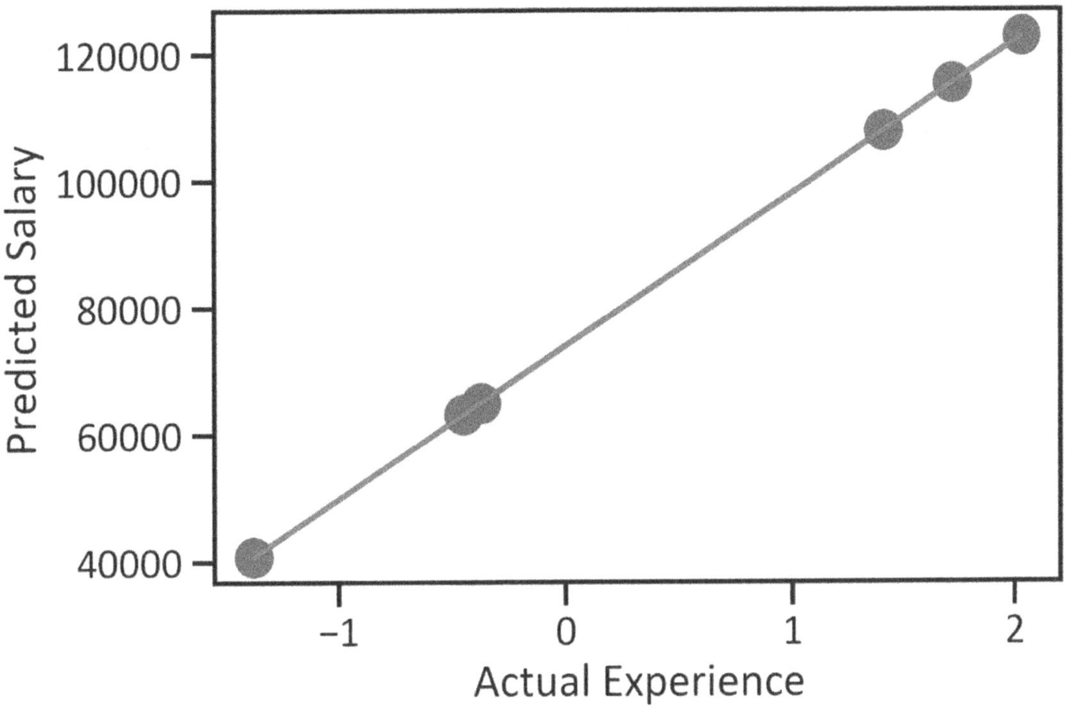

Figure 1-14. *Test data*

Figure 1-14 affirms a perfect fit model. As mentioned, a perfect model is rare in the actual world because of variability.

Actual Values and Predicted Values

Model evaluation involves comparing the actual values and the predicted values of a dependent variable. Listing 1-36 returns the difference in actual values of employees' salaries against predicted values of employees' salaries (see Figure 1-15).

Listing 1-36. Actual Values and Predicted Values

```
plt.scatter(y_test,y_pred,s=200)
plt.axhline(color="red")
plt.xlabel("Actual Salary")
plt.ylabel("Predicted Salary")
plt.show()
```

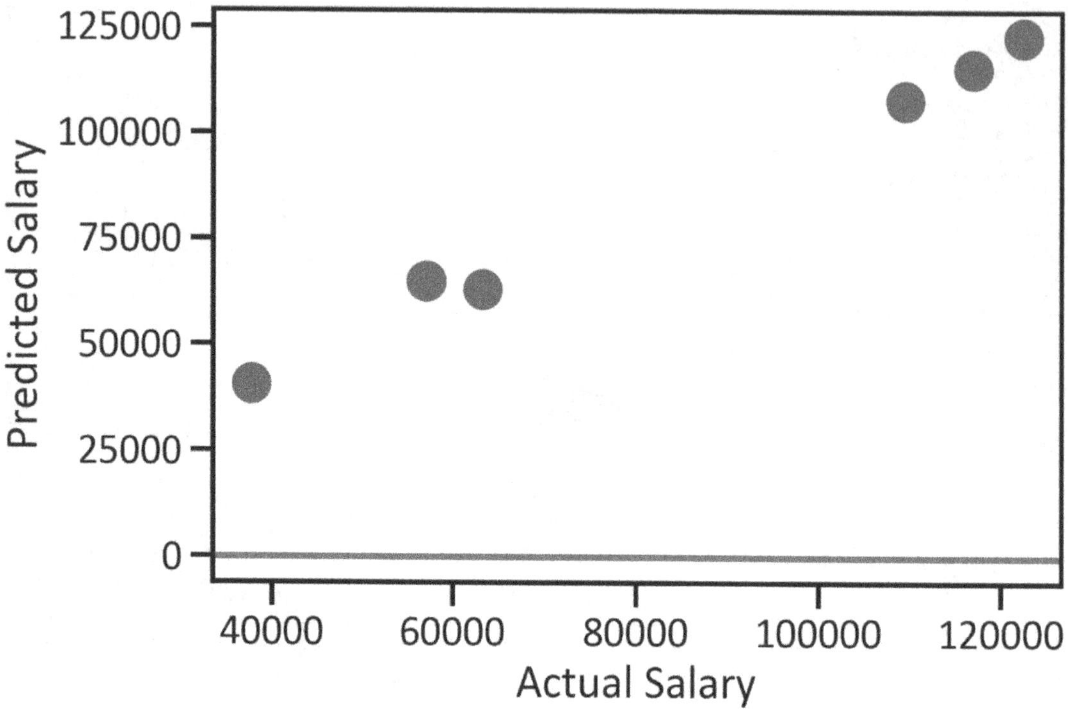

Figure 1-15. *Actual values and predicted values*

There are notable differences between the actual values of employees' salaries and predicted values of employees' salaries, but those differences are not large.

Diagnose Residuals

A residual represents the difference between the actual values and the predicted values. Residual diagnosis involves discovering behavioral patterns of errors that a model makes. There are several tools that we can use to check whether a model fulfils the assumption. The easiest way of analyzing the residuals involves looking at their mean value. If the mean value of the residuals is close to 0, then the model is a good fit.

To graphically represent residuals and make sense of the behavioral patterns of errors, we can use plots such as the autocorrelation function plot, partial autocorrelation function plot, fitted values and residuals values plot, leverage values and residual values plot, fitted values and studentized residual values, leverage and studentized residual values (also recognized as Cook's D influence plot), and the normal probability (or normal Q-Q) plot. These plots are useful for detecting abnormalities in residuals.

If there is a pattern or shape in the residuals, then the model violates a regression assumption. There are three main patterns, namely, *fan-shaped pattern,* which happens when there is an increase of variability with predicted values (error term not affected by the value of an independent variable); *parabolic pattern,* which happens when predicted values are relative or there are multiplicative errors; and *double bow pattern,* which happens when an independent variable is a proportion or percentage. There are other factors that influence behavioral patterns of residuals such as the distribution of the data and visible presence of outliers. It is important to ensure that we feed the model quality data.

Evaluate Residuals Using Statsmodels

Listing 1-37 shows the residuals diagnosis metrics, and Listing 1-38 shows the box plot.

Listing 1-37. Residuals Diagnosis Metrics

```
model_residual = model.resid
model_fitted = model.fittedvalues
model_leverage = model.get_influence().hat_matrix_diag
model_norm_residual = model.get_influence().resid_studentized_internal
model_norm_residual_ab_sqrt = np.sqrt(np.abs(model_norm_residual))
```

Figure 1-16 shows the dispersion of the residuals.

Listing 1-38. Residuals Box Plot

```
model_residual = pd.DataFrame(model_residual)
model_residual.columns = ["Residuals"]
sns.boxplot(model_residual["Residuals"])
plt.xlabel("Residuals")
plt.show()
```

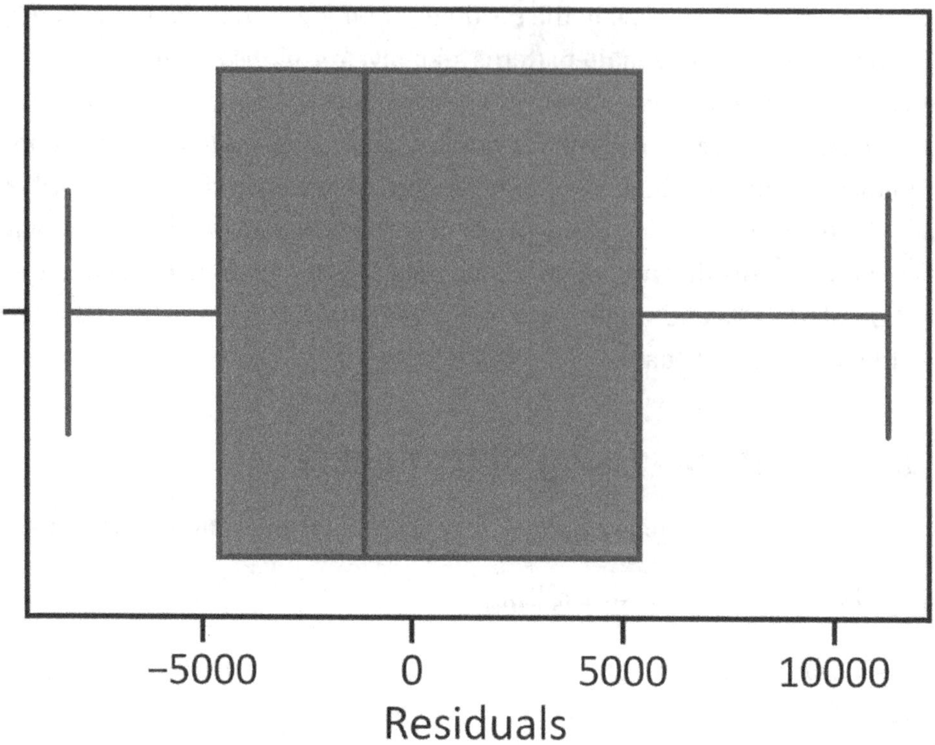

Figure 1-16. *Residuals box plot*

Figure 1-16 shows that the mean value of residuals is close to 0, but the closeness is not extreme. The residuals spread symmetrically around the true value (they are well-behaved).

Normal Q-Q

Listing 1-39 plots the normal Q-Q. Figure 1-17 determines whether data comes from a normal theoretical distribution. It verifies the assumption of normality by comparing two probability distributions (theoretical quantiles on the x-axis and values of the sample quantiles on the y-axis).

Listing 1-39. Normal Q-Q

```
fig, ax = plt.subplots()
fig = sm.graphics.qqplot(model_residual,ax=ax,line="45",fit=True,
dist=stats.norm)
plt.show()
```

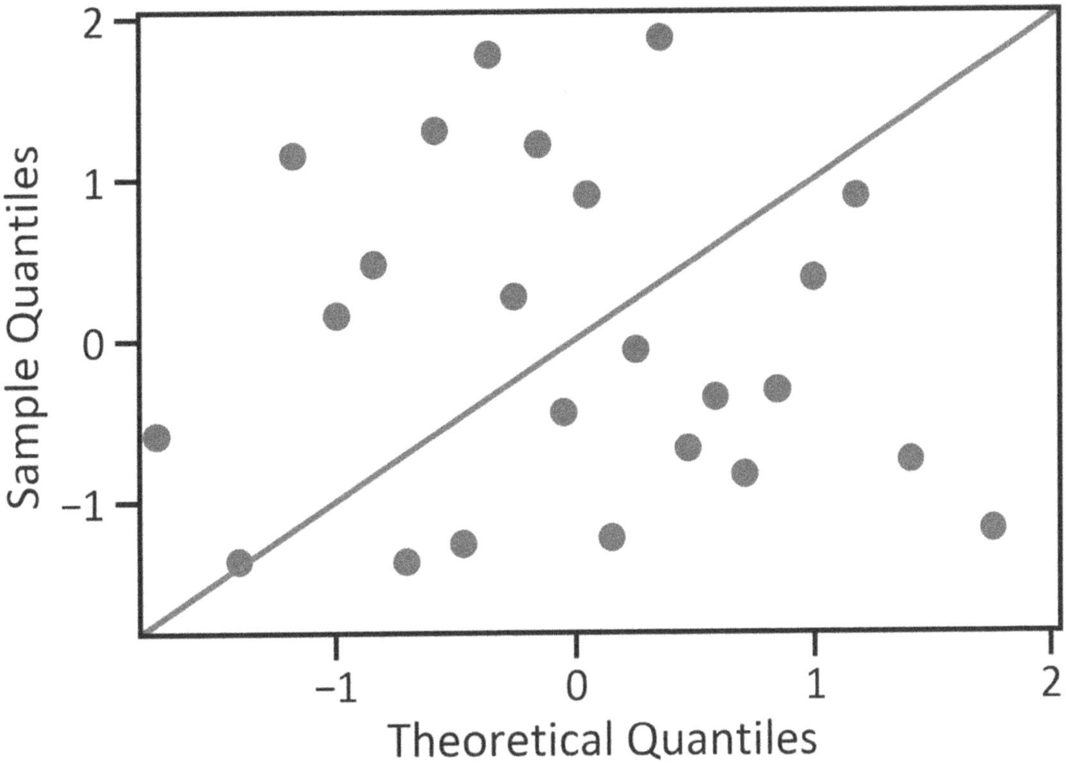

Figure 1-17. *Normal Q-Q*

Figure 1-17 shows that data points do not form a straight line. They deviate away from the straight line. It also shows outliers in the data.

Cook's D Influence

Listing 1-40 plots the Cook's D influence. Figure 1-18 detects outliers that pull a regression function toward itself. It uses a cutoff value to determine outliers. A cutoff represents a value at which we decide whether an observation is an outlier. We estimate the cutoff using 4/n-k-1, where *n* is the sample size and *k* is the number of outliers for assessing the performances.

Listing 1-40. Cook's D Influence

```
fig, ax = plt.subplots()
fig = sm.graphics.influence_plot(model,ax=ax,criterion="cooks")
plt.show()
```

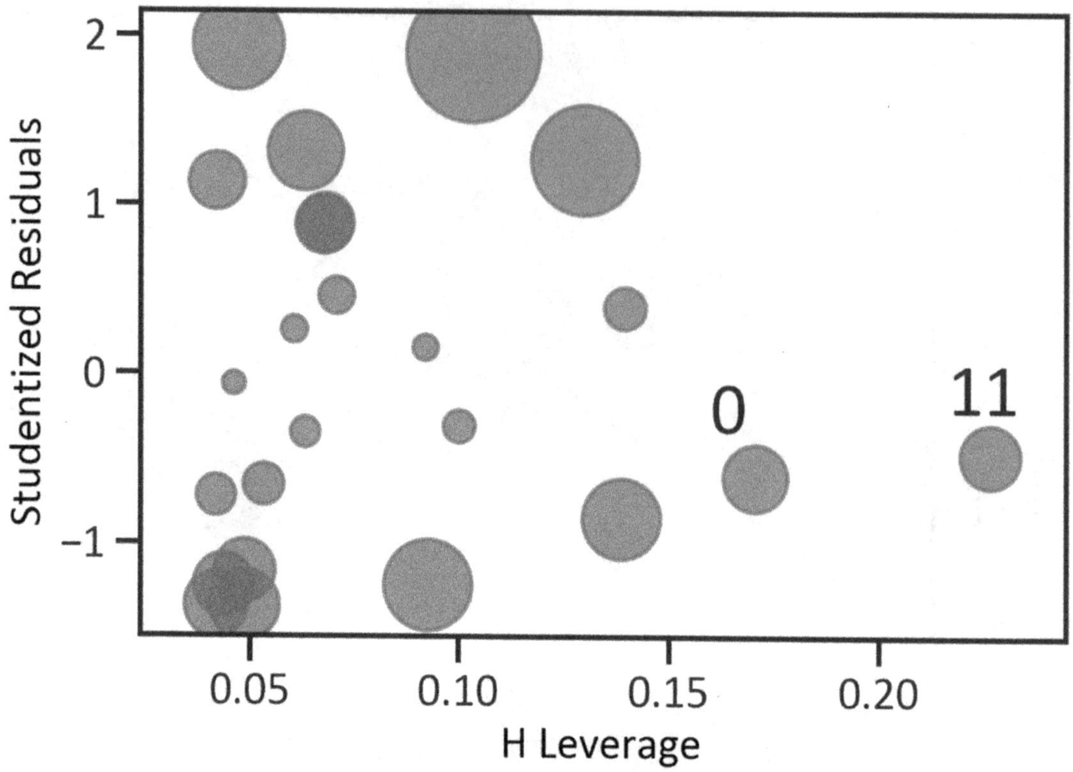

Figure 1-18. *Cook's D influence*

Figure 1-18 shows influential data points (a minor point at observation 0 and an extreme one at observation 11). Removing them significantly improves the predictive power of the model.

Fitted Values and Residual Values

Listing 1-41 plots fitted values and residual values. Figure 1-19 compares fitted values (predicted responses) and against residuals (the difference between actual values and predicted values). It checks for unequal variances, nonlinearity, and outliers.

Listing 1-41. Fitted Values and Residual Values

```
plt.scatter(model_fitted,model_residual,s=200)
plt.xlabel("Fitted Salary")
plt.ylabel("Residual Salary")
plt.show()
```

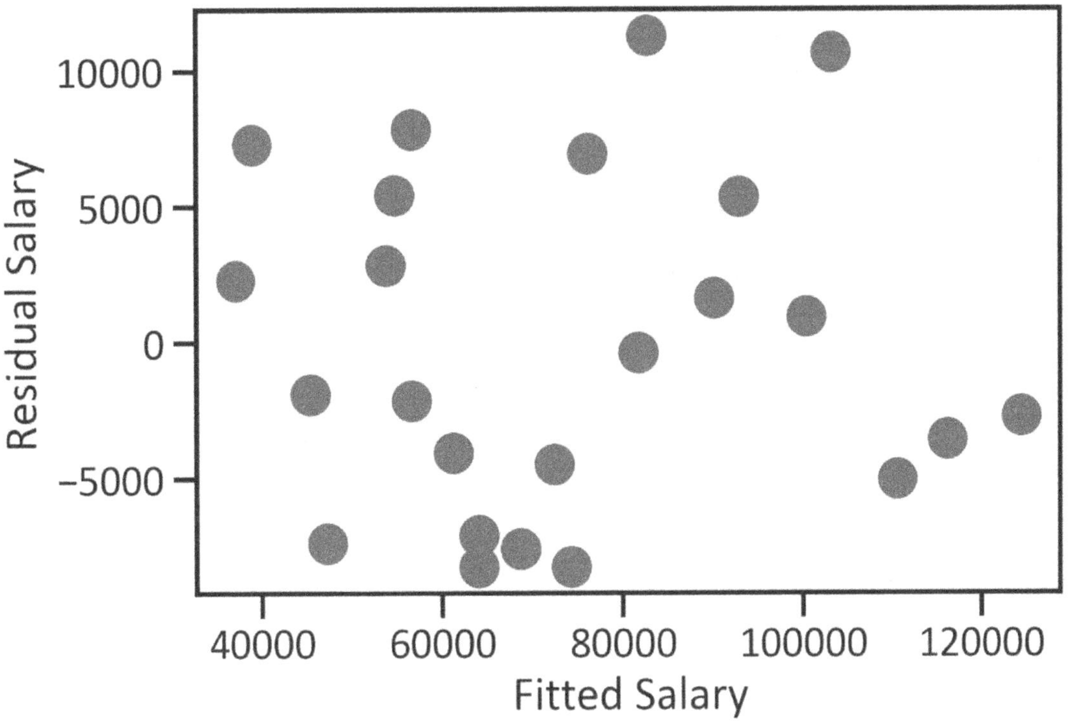

Figure 1-19. *Fitted values and residual values*

Figure 1-19 shows the characteristics of a well-behaved fitted salary and residual salary plot. There are no large residuals. Also, residuals bounce randomly around zero. The model satisfies the assumption of linearity.

Leverage Values and Residual Values

Listing 1-42 plots leverage values (how far way *x* is from the mean of y) and residuals (see Figure 1-20).

Listing 1-42. Leverage Values and Residual Values

```
plt.scatter(model_leverage,model_residual,s=200)
plt.xlabel("Leverage Salary")
plt.ylabel("Residual Salary")
plt.show()
```

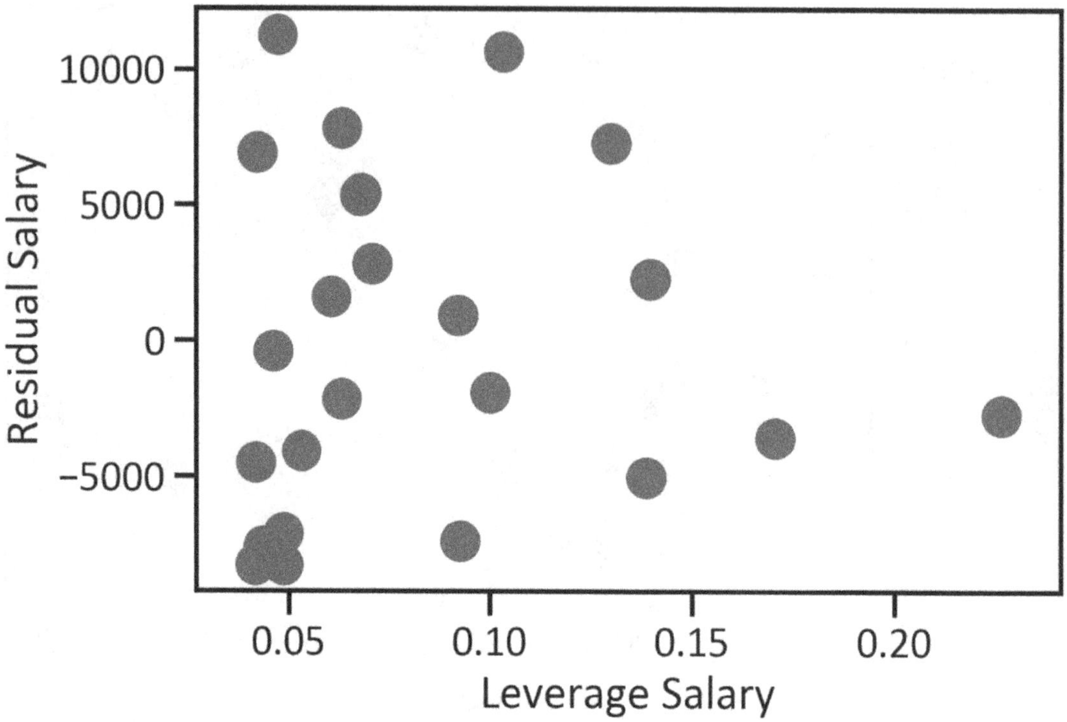

Figure 1-20. *Leverage values and residual values*

Figure 1-2 suggests that there is one high leverage data point that contributes to prediction inaccuracy.

Fitted Values and Studentized Residual Values

Listing 1-43 returns predicted responses and standardized deleted residuals (see Figure 1-21). Studentized residuals equal to or greater than 3 are influential points.

Listing 1-43. Fitted Values and Studentized Residual Values

```
plt.scatter(model_leverage,model_norm_residual,s=200)
plt.xlabel("Leverage Salary")
plt.ylabel("Studentized Residual Salary")
plt.show()
```

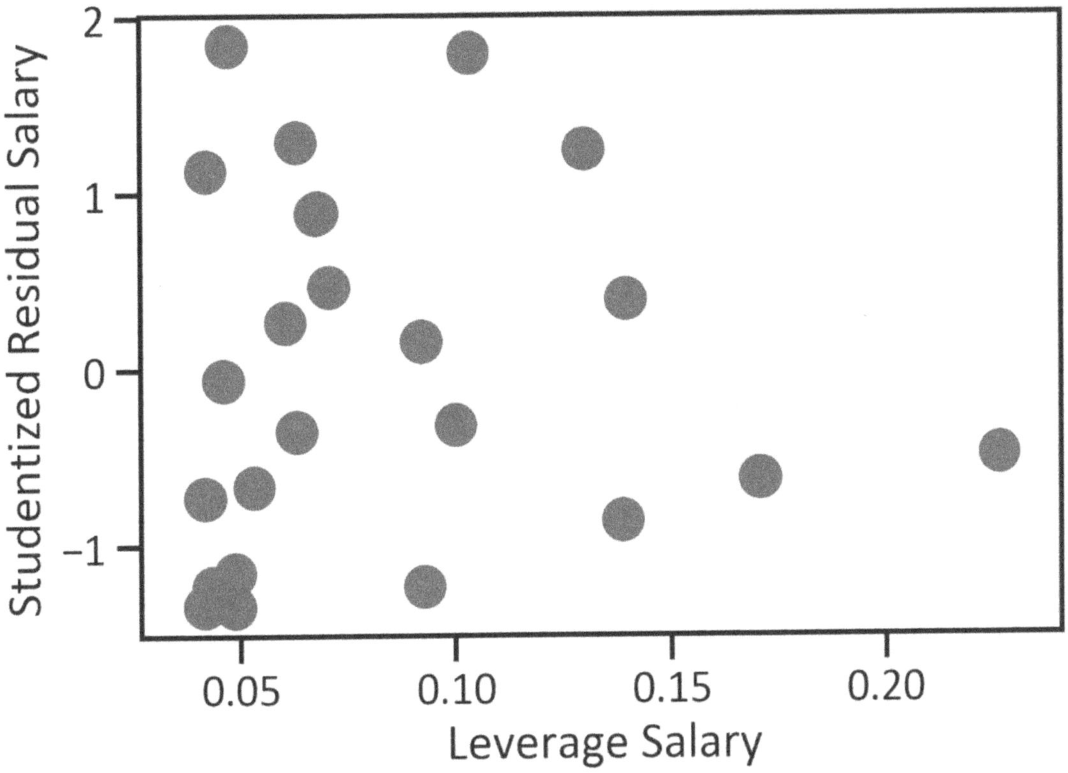

Figure 1-21. *Fitted values and studentized residual values*

Figure 1-21 confirms random distribution (residuals do not follow a trend or pattern).

Leverage Values and Studentized Residual Values

Another way of identifying influential data points involves comparing how far data points of an independent are from other data points against standardized deleted residuals. See Listing 1-44.

Listing 1-44. Fitted Values and Studentized Residual Values

```
plt.scatter(model_leverage,model_norm_residual,s=200)
plt.xlabel("Leverage Salary")
plt.ylabel("Studentized Residual Salary")
plt.show()
```

Figure 1-22 shows characteristics of well-behaved residuals, but there is one extreme outlier.

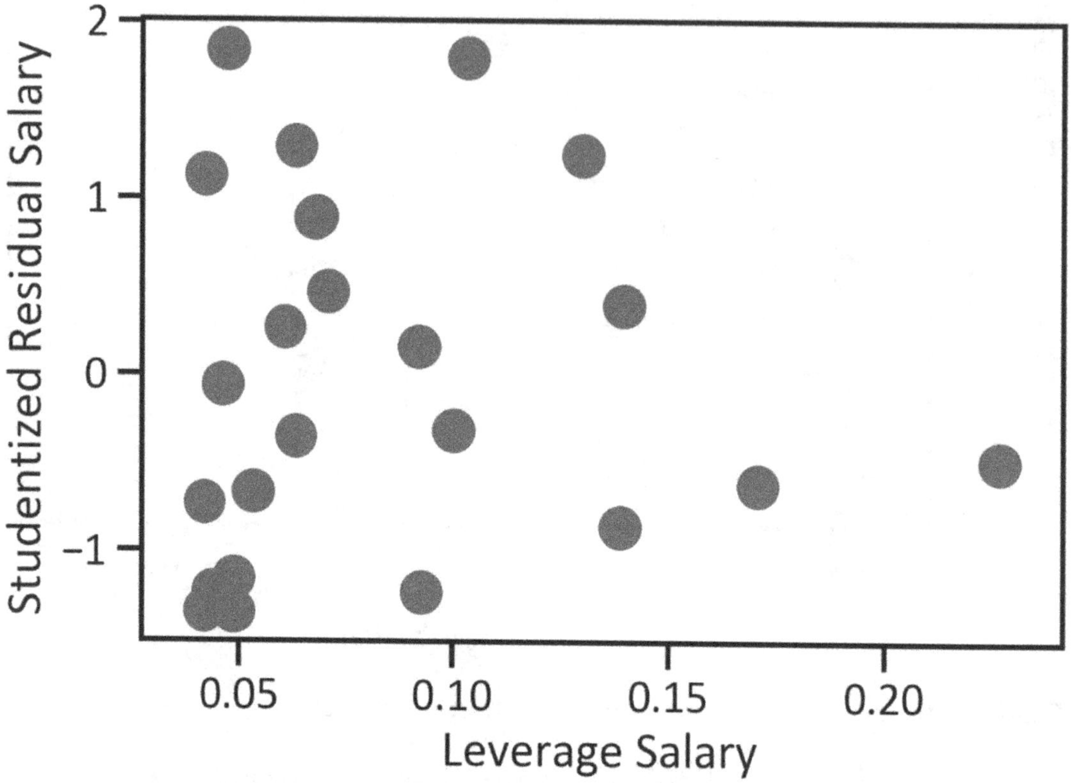

Figure 1-22. *Fitted values and studentized residual values*

Evaluate Residuals Using SciKit-Learn

The following section investigates the extent to which to residuals correlate with themselves using the autocorrelation function and the partial autocorrelation function.

Autocorrelation Function

Autocorrelation represents the extent to which data correlates with itself, as opposed to some other data (see Listing 1-45). Figure 1-23 determines whether there is autocorrelation between residuals in previous lags. It has two axes, with lags on the x-axis and with the autocorrelation function of residuals on the y-axis.

Listing 1-45. Autocorrelation Function

```
from statsmodels.graphics.tsaplots import plot_acf
residuallm = y_test - y_pred
plot_acf(residuallm)
plt.xlabel("Lag")
plt.ylabel("ACF")
plt.show()
```

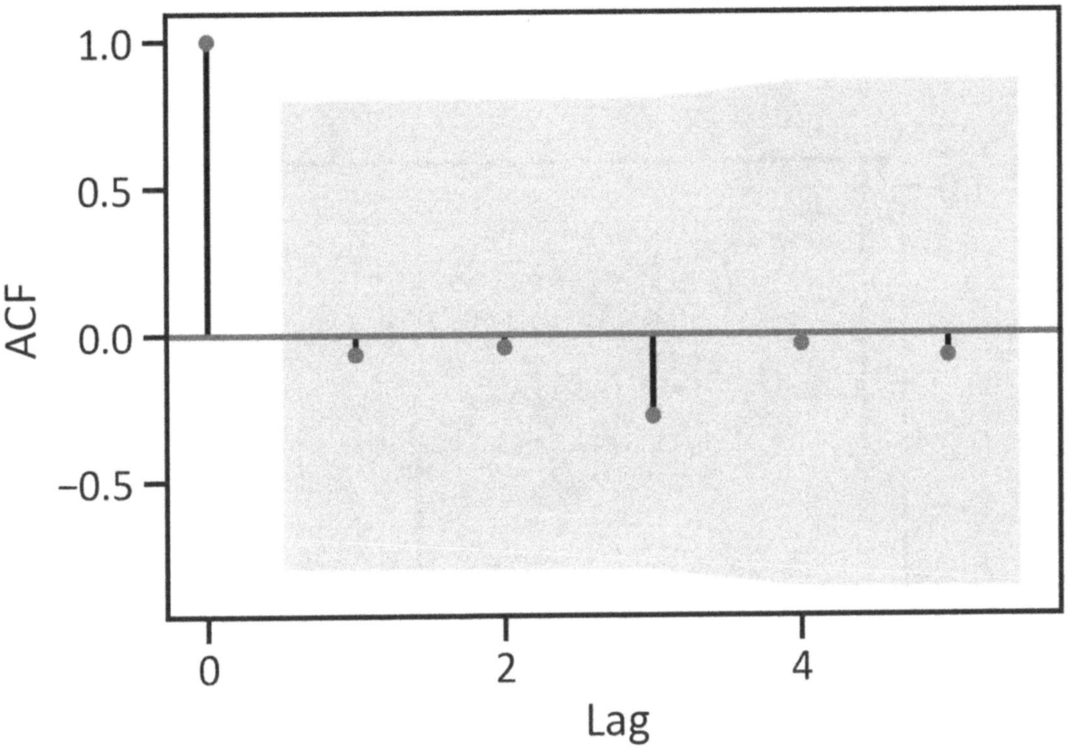

Figure 1-23. *Autocorrelation function plot*

Figure 1-23 shows that at lag 0, the correlation is 1 (the data correlates with itself). There is a slight negative correlation at lags 1, 2, 4, and 5 and a moderate negative correlation at lag 3.

Partial Autocorrelation Function

Listing 1-46 plots the partial correlation of coefficients not explained at low-level lags (see Figure 1-24).

Listing 1-46. Partial Autocorrelation Function

```
from statsmodels.graphics.tsaplots import plot_pacf
plot_pacf(residualslm)
plt.xlabel("Lag")
plt.ylabel("PACF")
plt.show()
```

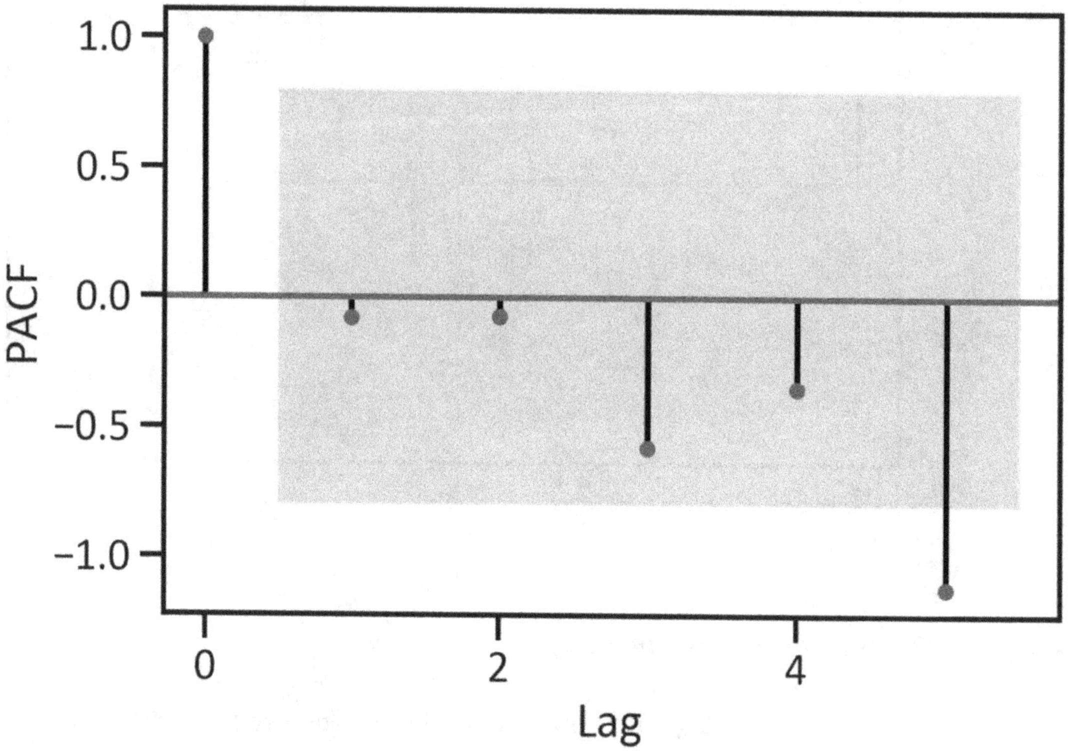

Figure 1-24. *Partial autocorrelation function*

Lag 0 can explain all higher-order autocorrelation.

Conclusion

This chapter covered the simple linear regression method and the least-squares model and its application. It examined whether there is a significant difference between employees' work experience and employees' salaries using two regressors (applying both Statsmodels and SciKit-Learn). The findings suggest there is a significant difference between employees' work experience and employees' salaries. There is linearity, but the regressors violate certain regression assumptions. They make marginal errors due to minor abnormalities in the data (this is common with a small sample). The subsequent chapter covers techniques for dealing with errors; it introduces the concept of errors due to bias and variance and how to solve the errors using advanced parametric methods such as ridge regression and lasso regression.

CHAPTER 2

Advanced Parametric Methods

The method covered in the previous chapter violated certain regression assumptions. It cannot capture noise, and as a result, it makes mistakes when predicting future instances. The most convenient way of combating this problem involves adding a penalty term to the equation.

This chapter introduces the novel concept of bias-variance trade-off, and it then covers regularized models like ridge regression, ridge with built-in cross-validation regression, and lasso regression. Last, it compares the performance of these models against that of the least-squares model. To tacitly understand regularizing models, you must first understand the concept of bias-variance trade-off. In the following section, we properly introduce the concept.

Concepts of Bias and Variance

Bias represents the closeness of estimates to the actual value, and *variance* represents how the data points vary for each realization of a model. Error due to bias occurs when a model class cannot fit the data. Developing a more expressive model class combats this problem. On the other hand, error due to variance represents the variability of a model's prediction on a data point. This occurs when a model class can fit the data, but cannot do so. Developing a less expressive model class combats this problem.

© Tshepo Chris Nokeri 2021
T. C. Nokeri, *Data Science Revealed*, https://doi.org/10.1007/978-1-4842-6870-4_2

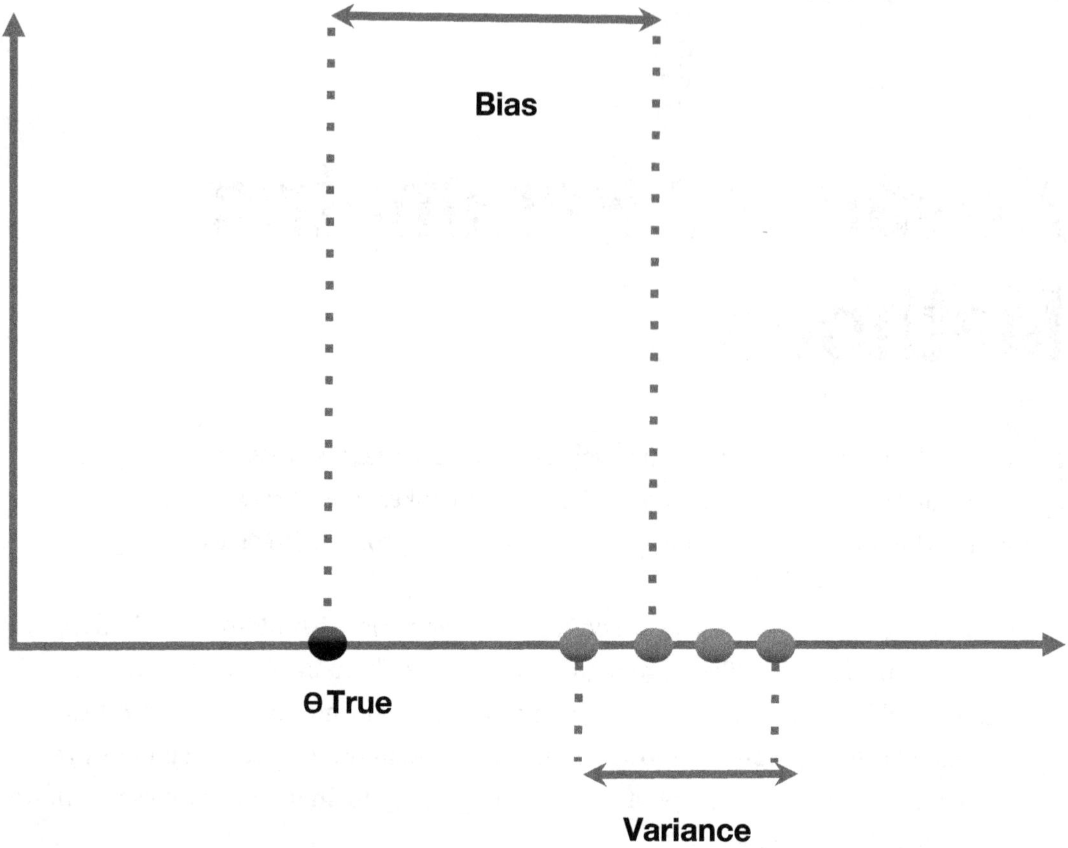

Figure 2-1. *Bias and variance graphic definition*

Figure 2-1 shows bias and variance.

Bias-Variance Trade-Off in Action

The bias-variance trade-off involves decreasing bias or variance at the expense of the other. We decrease variance to increase bias when there are a few variables in the data, there is a highly regularized model, the decision tree model has extremely pruned decision trees, and the K-nearest neighbor model has a large k. Figure 2-2 shows an example of low bias (or high variance).

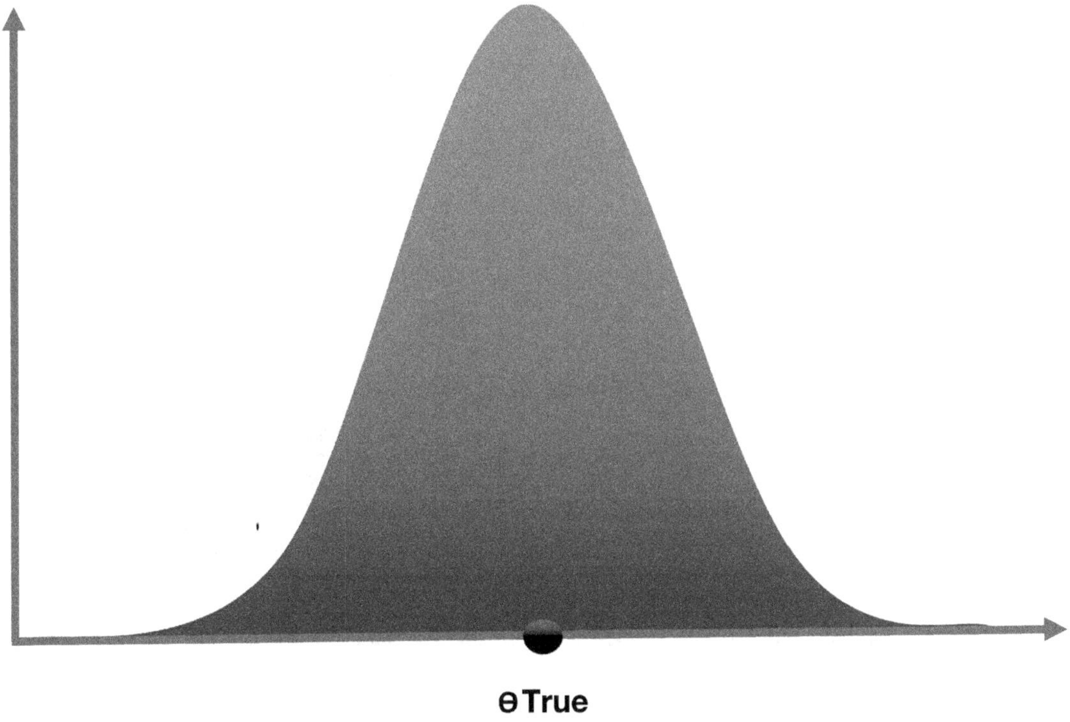

Figure 2-2. *Low bias/high variance*

We decrease bias to increase variance when there are several variables in the data, the model is unregularized, the decision tree model is unpruned, or the K-nearest neighbor model has a small k. Figure 2-3 shows an example of high bias (or low variance).

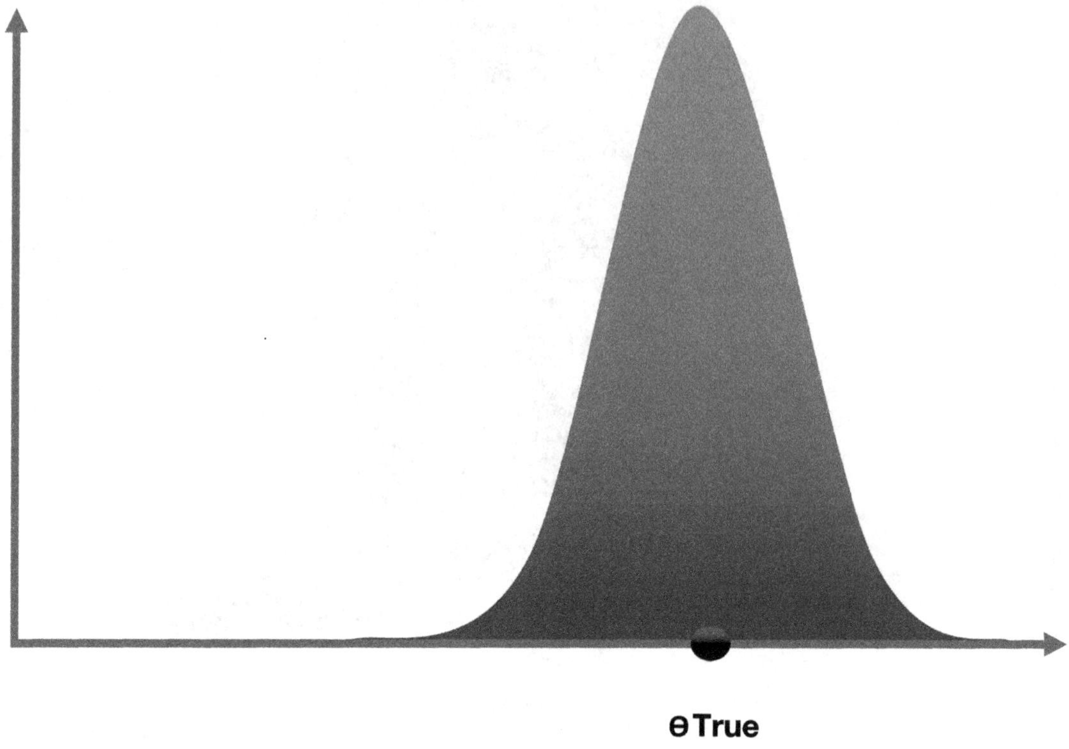

Figure 2-3. *High bias/low variance*

Bias-variance trade-off is about addressing the problem of underfitting (when there is a low bias error term) and overfitting (when there is high variance). We decrease bias and variance based on the complexity of the model. Several parameters in a model increase the complexity of the model and variance and decrease bias.

Ridge Regression

We also recognize ridge regression as Tikhonov or L2 regularization. This model addresses the problem of multicollinearity by minimizing variables by a small k along each eigenvector to ensure that there is a narrow spread in axes. Similar to the least squares regressor, it assumes linearity and normality in the data. The major difference between the two models is that the ridge regressor assumes that after normalizing the data, coefficients are small. It also assumes that as the value of k increases, coefficients with multicollinearity alter their behavior. Given this assumption, the model sharply decreases coefficients to zero. At most, the value of k of the ridge regressor is less than the value of k of the least squares model.

It shrinks the estimates to stabilize variability by adding bias (introducing a penalty and controlling for shrinkage to the least squares function). It prevents the values of the independent variable from exploding, reduces standard errors, and provides more reliable estimates. This procedure minimizes the penalized sum of residuals and carefully controls the considerable amount of shrinkage in a regression model. It changes the cost function by adding a penalty term (a term relative to the sum of squares of the coefficients). A constant penalty term ranges from 0 to 1. It controls estimates, β, and as a result, high variance is more likely to occur. The least squares model does not control the loss function of the ridge model. Equation 2-1 expresses a standard ridge formula.

$$\beta^{ridge} = (X'X + \lambda_2)^{-1} X'y \qquad \text{(Equation 2-1)}$$

Here, λ_2 represents the ridge penalty term responsible for penalizing the squared regression coefficient. The penalty term controls the size of the coefficients and the amount of regularization.

RidgeCV

RidgeCV stands for "ridge regression with built-in cross-validation" (it weighs the prediction errors of a model to validate each data point as an out-of-sample prediction). CV methods include the K-fold, leave-one-out, and generalized cross-validation (GCV). By default, the ridgeCV regressor in SciKit-Learn uses GCV. See Equation 2-2.

$$GCV = \frac{1}{n}\sum_{i=1}^{n}\left(\frac{y_i - \hat{f}(x_i)}{1 - \frac{tr(S)}{n}}\right) \qquad \text{(Equation 2-2)}$$

Recall that *tr(S)* is the effective number of parameters.

It alters model behavior by partitioning data, finding optimal penalty terms, and training and validating complementary regressors.

Lasso Regression

Lasso stands for "least absolute shrinkage selector operator" (also known as L1 regularization). It applies the shrinkage technique (centering the data to its central point, such as centering the data to the mean value). This procedure produces sparse models (models with a reduced numbers of parameters). We use it for both variable selection and regularization. Like ridge regression, lasso regression boosts a regressor's predictive power by including a penalty term. They differ from the penalty term: lasso regression uses L1 to penalize the sum of residuals, while ridge regression uses the L2 penalty term. It assumes that as the penalty term increases, more coefficients were set to zero, which results in fewer variables being selected. L1 regularization uses nonzero coefficients and coerces the sum of absolute values of coefficient (sets 0 coefficients and adds interpretability). See Equation 2-3.

$$\beta^{lasso} = (X'X)^{-1} X'y - \frac{\lambda_1}{2} w \qquad \text{(Equation 2-3)}$$

λ1 represents the lasso penalty, penalizing the sum of the absolute values of the regression coefficients.

Tunable Hyperparameters

If you do not want to use default parameters to complete your model, you can change the hyperparameters accordingly. Table 2-1 highlights tunable hyperparameters.

Table 2-1. *Tunable Hyperparameters*

Parameters	Description
alpha	Determines the regularization strength. It is a constant that multiplies the term. One must be cautious when selecting alpha. A small value of alpha specifies weak regularization, and a large value specifies strong regularization. As alpha increases, the parameters become smaller. If alpha increases, coefficients approach 0 and under-fit the data. The default value is 1.
fit_intercept	Determines whether the model must estimate the intercept.
normalize	Determines whether we must normalize the independent variables.
copy_X	Determines whether we must copy the independent variable.
tol	Determines the precision. The default value is 0.003.
max_iter	Determines the maximum number of iterations.
tol	Determines the tolerance for the optimization
warm_state	Determines whether to initialize a solution or erase the preceding solution.

Develop the Models

The example data was obtained from GitHub.[1] It compromises a few categorical independent variables. Listing 2-1 transforms the categorical variable into numeric using the `LabelEncoder()` method. For example, we transform `['VS2' 'VS1' 'VVS1' 'VVS2' 'IF']` to `[0,1,2,34]`. Listing 2-2 creates an x and y array and thereafter splits and normalizes the data.

Listing 2-1. Convert Variables to Numeric

```
from sklearn.preprocessing import LabelEncoder
categorical_features = ['colour', 'clarity']
le = LabelEncoder()
```

[1] https://vincentarelbundock.github.io/Rdatasets/csv/Ecdat/Diamond.csv

```
for i in range(2):
    new = le.fit_transform(df[categorical_features[i]])
    df[categorical_features[i]] = new
df["certification"] = pd.get_dummies(df["certification"])
print(df.clarity.unique())
print(df.colour.unique())
print(df.certification.unique())
```

```
[2 1 3 4 0]
[0 1 3 2 4 5]
[1 0]
```

Listing 2-1

Last, Listing 2-3 trains all the models.

Listing 2-2. Data Preprocessing

```
from sklearn.preprocessing import StandardScaler
from sklearn.model_selection import train_test_split
x = df.iloc[::,0:4]
y = df.iloc[::,-1]
x_train, x_test, y_train, y_test = train_test_split(x,y,test_size=0.2,
random_state=0)
scaler = StandardScaler()
x_train = scaler.fit_transform(x_train)
x_test = scaler.transform(x_test)
```

Listing 2-3. Develop Models

```
from sklearn.linear_model import LinearRegression, Ridge, RidgeCV, RidgeCV
lm = LinearRegression()
lm.fit(x_train,y_train)
ridge = Ridge()
ridge.fit(x_train, y_train)
ridgecv = RidgeCV()
ridgecv.fit(x_train, y_train)
lasso = RidgeCV()
lasso.fit(x_train,y_train)
```

Evaluate the Models

Table 2-2 summarizes the performance of all regressors using primary evaluation metrics.

Table 2-2. *Model Evaluation Results of All Models*

	MAE	MSE	RMSE	R^2	Explained variance score
OLS	550.512799	528211.610638	726.781680	0.942619	0.944409
Ridge	548.303122	526871.168399	725.858918	0.942764	0.944672
Ridge CV	550.291000	528056.270472	726.674804	0.942636	0.944437
Lasso	550.291000	528056.270472	726.674804	0.942636	0.944437

Table 2-2 highlights that all regressors explain more than 94% of the variability in the data. However, the least-squares regressor is the weakest contender, with the lowest R-squared score and discrepancies. In addition, the ridge regressor has the highest discrepancy between predicted values and actual values.

Conclusion

This chapter explored the different models for regularizing estimates of the least squares model such as ridge, ridge with built-in cross-validation, and lasso. These models sufficiently dealt with error due to bias and error due to variance. After testing their performance, we found that regularizing or shrinking estimates of a mode improves the general performance. The ridge regressor model is a delicate contender. In a case where both regularizing models do not improve the predictive power of the least squares model, use the ElasticNet model, a viable hybrid of ridge regression and lasso regression.

CHAPTER 3

Time-Series Analysis

This introduces a complex regression method, called *time-series analysis*. Similar to the regression method, the time series makes strong assumptions about the underlying structure of the data. We recognize time-series analysis as a parametric method because it deals with a continuous variable. Remember, we do not limit series analysis to continuous variables; we can also analyze time-dependent categorical variables.

It introduces empirical tests for testing assumptions, and we then introduce the standard techniques for analyzing a time series like seasonal decomposition and smoothing. After that, it shows you how to estimate the rate of return and how to run the minimum and maximum. Last, it covers SARIMAX model development and evaluation.

What Is Time-Series Analysis?

Time-series analysis enables us to recognize apparent trends and consistent patterns in a series. It captures the movement of a variable across time. In continuous time-series analysis, we continuously examine a variable. We define a series at a specific point and not at each point. For instance, we will use closing prices on the foreign exchange (there is one closing price per trading trade). We obtained the example data from Yahoo (it is series data about the price of the USD/ZAR currency pair from July 3, 2010, to July 4, 2020).

Time-Series Assumptions

A time-series analysis model must fulfill the following assumptions:

- More than 50 independent data points
- No missing values
- No outliers in the series

© Tshepo Chris Nokeri 2021
T. C. Nokeri, *Data Science Revealed*, https://doi.org/10.1007/978-1-4842-6870-4_3

- Nonstationary
- Absence of white noise
- Correlation between variables with themselves

Types of Time-Series Models

We typically analyze a series by smoothing or moving averages (MA) using the box-Jenkins autoregressive (AR) and MA model, which is a combination of these methods called the ARMA model. When a series convincingly shows a pattern, we use a relaxed version of the model recognized as the autoregressive integrated moving averages (ARIMA) model. We can include a seasonal component using the seasonal ARIMA or SARIMAX model. These methods describe a series in terms of its linear structure. Sometimes, there is a nonlinear structure. When dealing with a nonlinear model, use the autoregressive conditionally heteroscedasticity (ARCH) model. When dealing with multiple time series, use the VARMA model.

The ARIMA Model

ARIMA (p, d, q) finds a key trend and reasonably predicts future values of the series, where p represents the order of the autoregressive model, d represents the degree of differencing, and q represents the order of the moving average model. The term ARIMA is threefold: AR (autoregressive) represents the linear combination of preceding values' influence, I (integrative) represents the random walk, and the MA (moving average) represents the linear combination of preceding errors.

Test for Stationary

The series is stationary when there are stochastic trends (random walk or unit root). This happens when there is uncertainty in the data. We recognize these trends when the series possesses elements of randomness. Their visible presence affects conclusions. Before training a time-series model, test whether a series is stationary. We express the hypothesis as follows:

> *Null hypothesis:* The series is stationary.
>
> *Alternative hypothesis:* The series is not stationary.

The Augmented Dickey-Fuller Test

The most reliable way to test whether a series is stationary is the augmented Dick-Fuller (ADF) test. It is a unit root test. When looking at the results, fiercely devote attention to the ADF F-statistics. If the F-statistics have a negative value, then there is sound evidence that there is a unit root. We express the hypothesis as follows:

> *Null hypothesis:* There is a unit root.
>
> *Alternative hypothesis:* There is no unit root.

If the p-value score exceeds 0.05, then we adamantly reject the null hypothesis in favor of the alternative hypothesis. There is sound evidence that there is no unit root.

Conduct an ADF Fuller Test

Listing 3-1 conducts an ADF Fuller test (see Table 3-1).

Listing 3-1. ADF Fuller Test

```
from statsmodels.tsa.stattools import adfuller
adfullerreport = adfuller(new_df["Close"])
adfullerreportdata = pd.DataFrame(adfullerreport[0:4],
                                  columns = ["Values"],
                                  index=["ADF F% statistics",
                                         "P-value",
                                         "No. of lags used",
                                         "No. of observations"])
Adfullerreportdata
```

Table 3-1. *ADF Test Statistics*

	Values
ADF F% statistics	-1.031626
P-value	0.741555
No. of lags used	0.000000
No. of observations	262.000000

As expected, the F-statistics have a negative value. The p-value exceeds 0.05. We reject the null hypothesis in favor of the alternative hypothesis. There is sound evidence that there is no unit root—the series is not stationary.

Test for White Noise

A series has white noise when there is an element of randomness. The apparent presence of randomness shows that there is no correlation between data points of a series—the autocorrelation is zero. White noise is an example of stationary. We express the hypothesis as follows:

Null hypothesis: There is white noise.

Alternative hypothesis: There is no white noise.

To check whether a series has white noise, look at the mean value. If the mean value is zero, then a series has white noise. If the mean value of a series is not zero, then there is no white noise. See Listing 3-2 and Figure 3-1.

Listing 3-2. Random White Noise

```
randval = np.random.randn(1000)
autocorrelation_plot(randval)
plt.show()
```

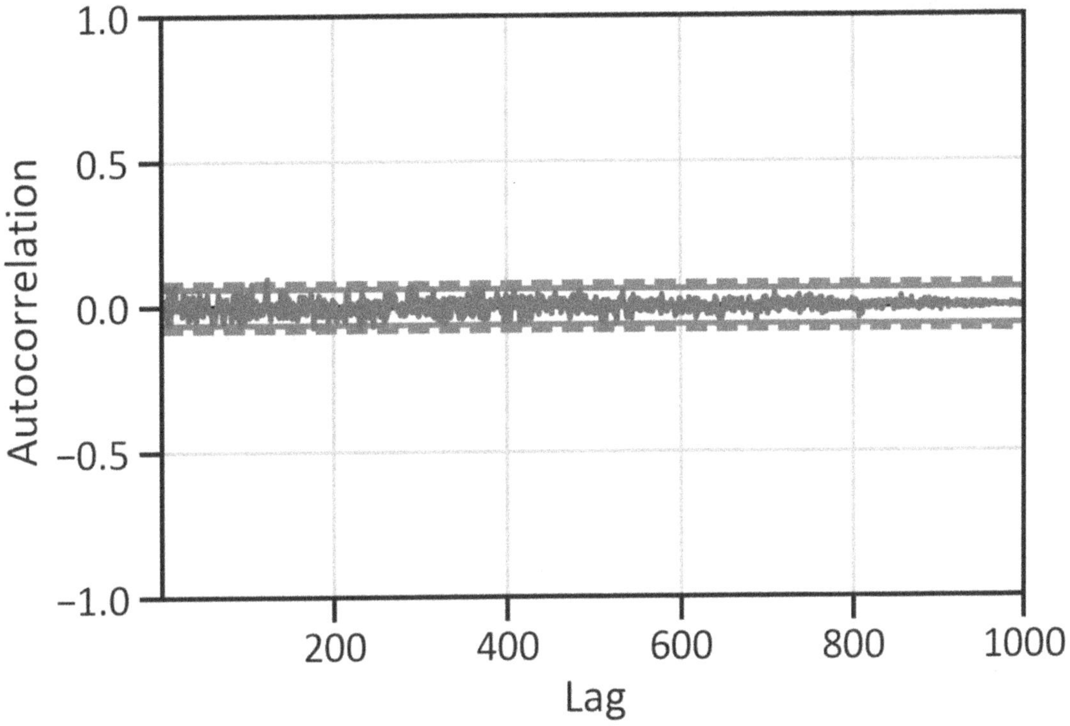

Figure 3-1. *Random white noise*

There are significant spikes above the 95% and 99% confidence interval.

Test for Correlation

Correlation estimates of the strength of a linear relationship between variables. In time-series analysis, we focus on the correlation between data points in a series. If you recall, we proceed under the assumption that the underlying structure of the data is linear. To check whether there is a correlation in a series, we do not use the Pearson correlation method. We verify the assumption using autocorrelation.

Listing 3-3 and Figure 3-2 show lagged variables. A lag is the difference in time between a recent data points and preceding data points. It verifies whether a series has an element of randomness. When we plot lags, we expect to find a pattern. If a lag plot shows a linear pattern, then there is a correlation between the two lagged variables. It also detects outliers in a time series. As you can see, there were no outliers detected.

Listing 3-3. Lag

```
lag_plot(df["Close"])
plt.show()
```

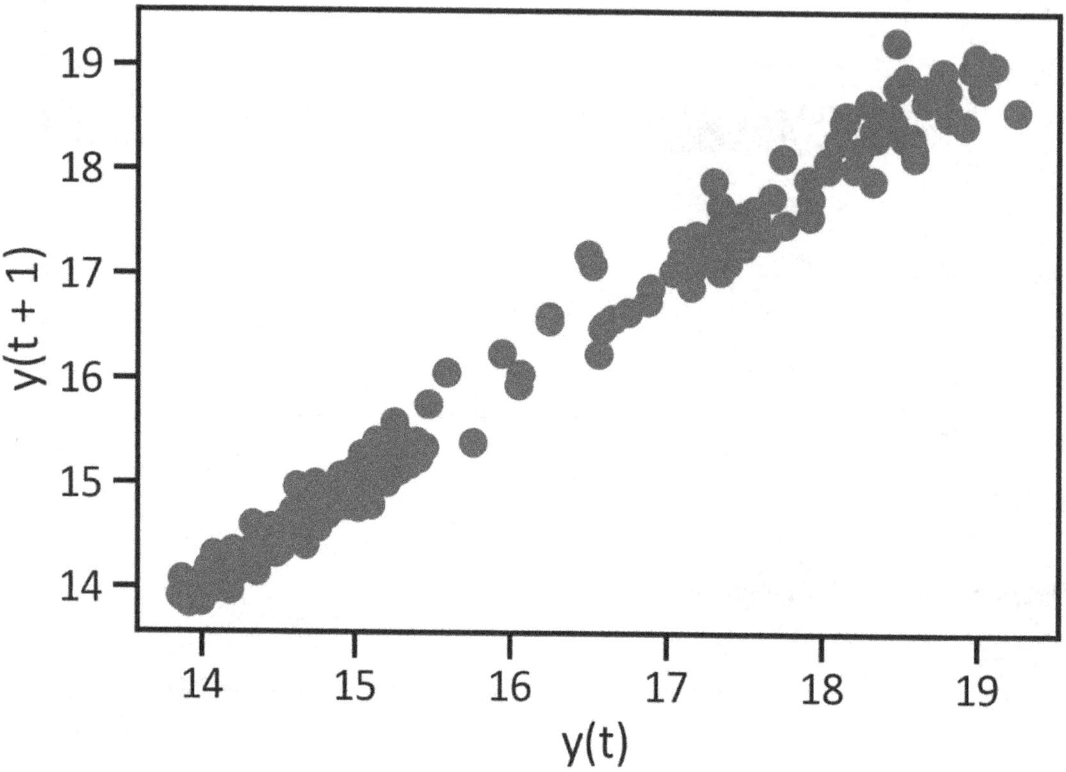

Figure 3-2. *Lag plot*

Figure 3-2 shows a linear pattern in the series (data points tightly aligned to an imaginary straight line). Listing 3-4 produces an autocorrelation plot (see Figure 3-3).

Listing 3-4. Autocorrelation

```
autocorrelation_plot(df["Close"])
plt.show()
```

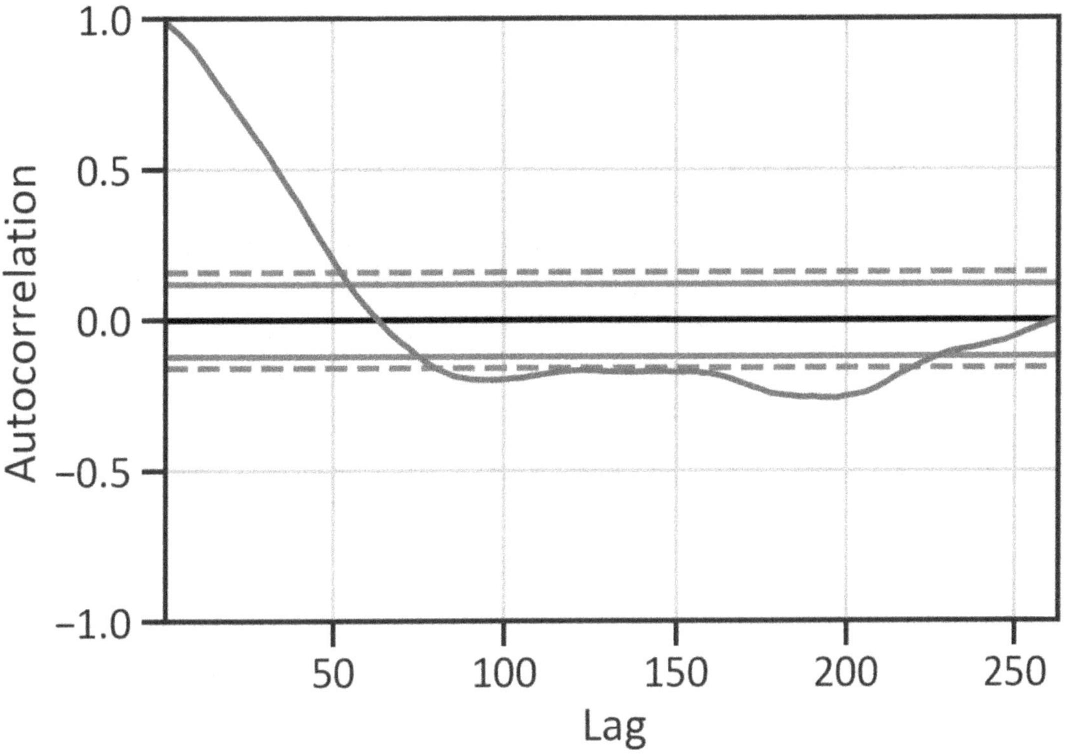

Figure 3-3. *Autocorrelation plot*

Figure 3-3 shows USD/ZAR daily returns where most of the spikes are not statistically significant.

Autocorrelation Function

Listing 3-5 shows the autocorrelation function (ACF) plot. Figure 3-4 considers the trend, seasonality, cyclic, and residual components.

Listing 3-5. ACF Plot

```
plot_acf(df["Close"])
plt.xlabel("Lag")
plt.ylabel("ACF")
plt.show
```

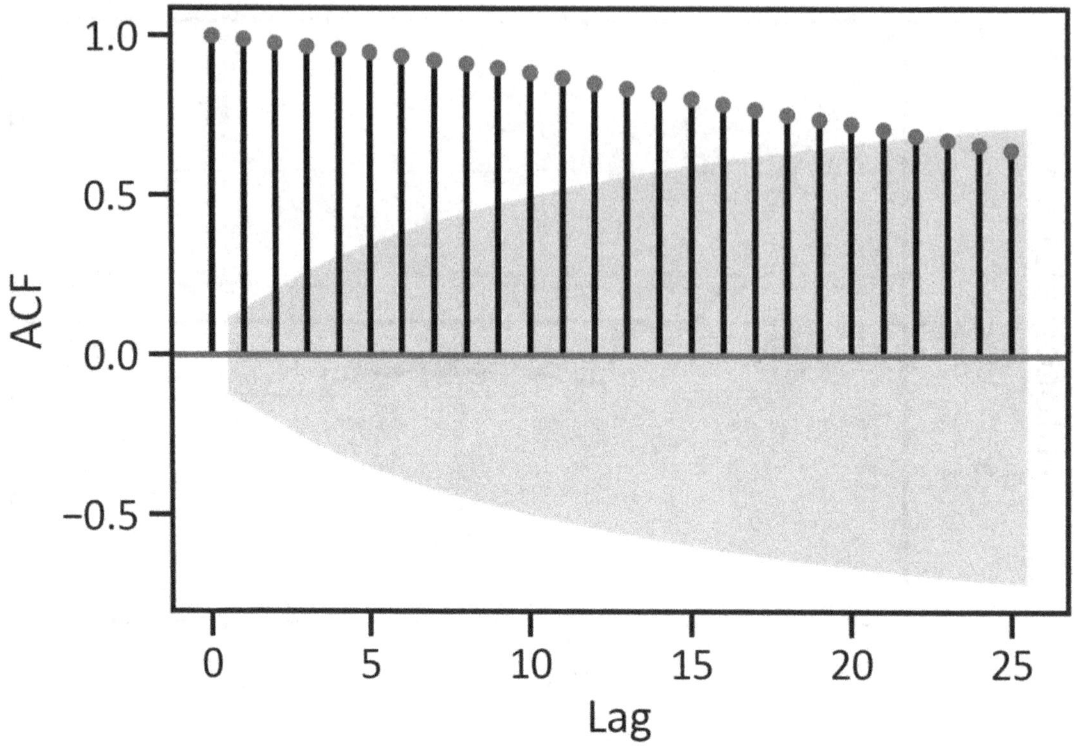

Figure 3-4. *ACF plot*

Partial Autocorrelation

Listing 3-6 shows partial autocorrelation (PACF). Figure 3-5 shows the partial correlation coefficient not explained at low-level lags. There is a significant spike at lag 1. The first lag explains all higher-order autocorrelation.

Listing 3-6. PACF

```
plot_pacf(df["Close"])
plt.xlabel("Lag")
plt.ylabel("PACF")
plt.show()
```

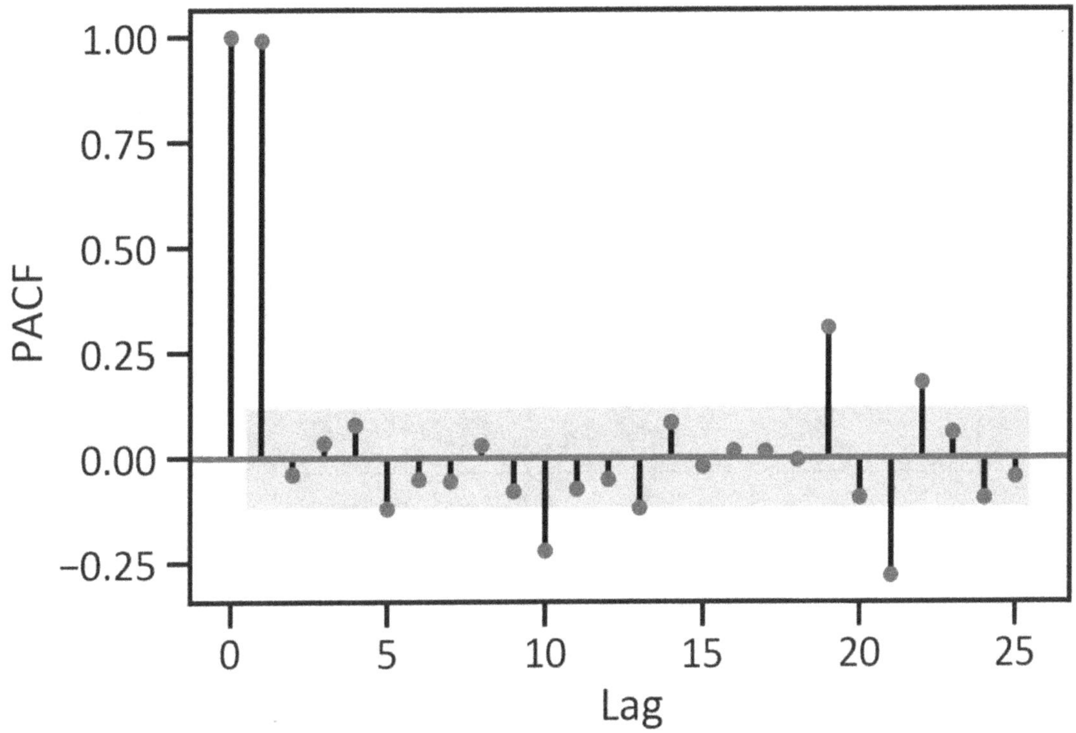

***Figure 3-5.** PACF plot*

Understand Trends, Seasonality, and Irregular Components

It is challenging to separate a trend from a cycle. The easiest way to make sense of time-series data is by breaking down data into different components. Seasonal decomposition involves breaking down a series into three components, namely, trend, cycles, seasonality, and irregular components. We commonly use an additive model, expressed as shown in Equation 3-1.

$$Y(t) = T(t)+C(t)+S(t) \times I(t) \quad \text{(Equation 3-1)}$$

Here, T(t) represents the trend value at period t, S(t) represents a seasonal value at period t, C(t) represents a cyclical value at time t, and I(t) represents an irregular (random) value at period t. Seasonal decomposition requires positive values. When there are zeros in a series, add 0.5 or 1 to the entire series. Listing 3-7 decomposes the series (see Figure 3-6).

Listing 3-7. Seasonal Decomposition

```
decompose = seasonal_decompose(df["Close"].interpolate(),freq=30)
decompose.plot()
plt.show()
```

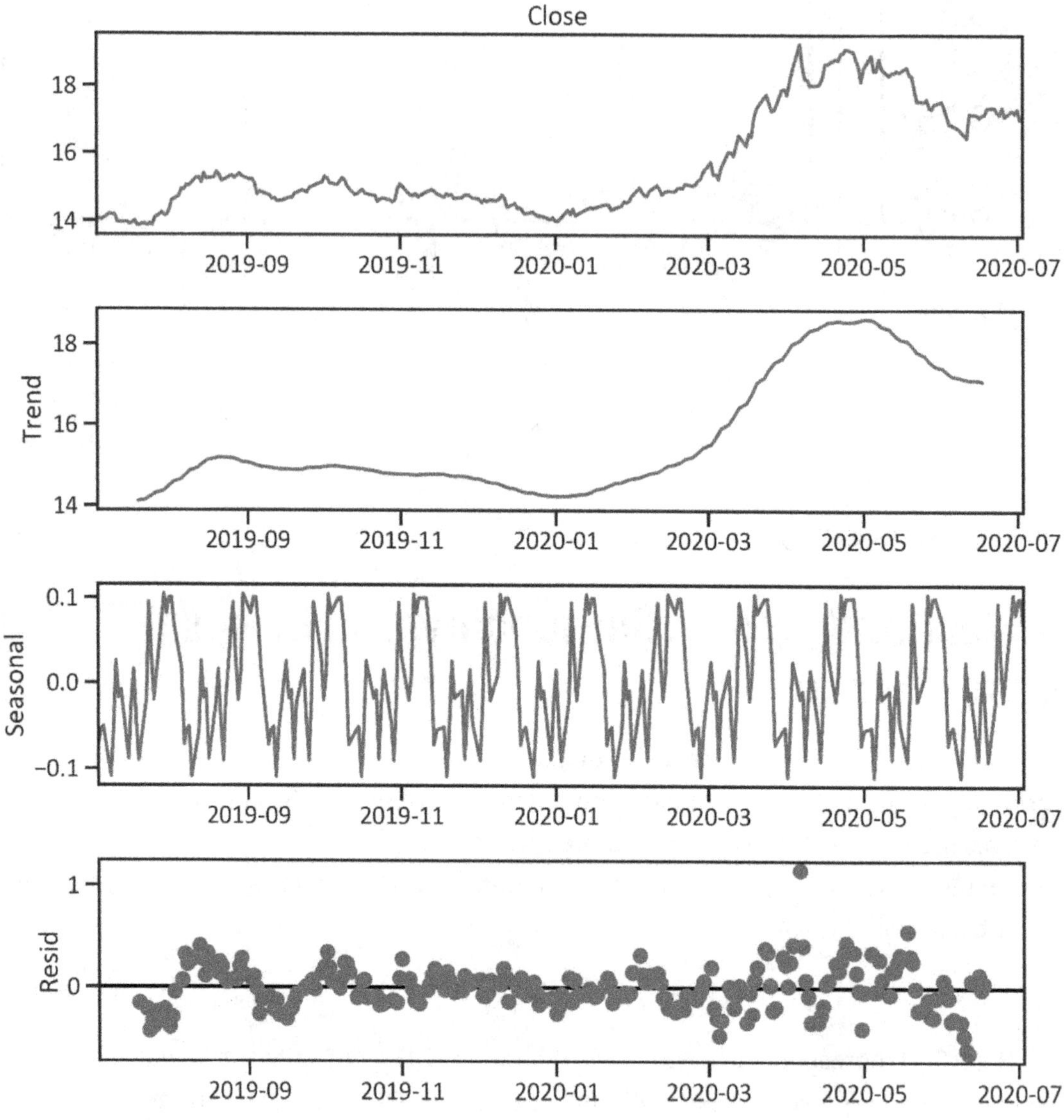

Figure 3-6. *Seasonal components*

Smoothing a Time Series Using Moving Average and Exponential Techniques

The most straightforward way to analyze a series involves using smoothing techniques. There are two primary smoothing techniques, namely, moving average smoothing and exponential smoothing.

Original Time Series

Listing 3-8 graphically represents the movement of a variable over time. Figure 3-7 has two axes: x-axis (horizontal) and y-axis (vertical), graphically denoted as (x,y). Time (an independent variable) is on the x-axis, and the dependent variable is on the y-axis.

Listing 3-8. Time Series

```
Original = new_df["Close"]
fig, ax = plt.subplots()
Original.plot(kind="line", color="navy")
plt.xlabel("Time")
plt.ylabel("Close")
plt.xticks(rotation=45)
plt.show()
```

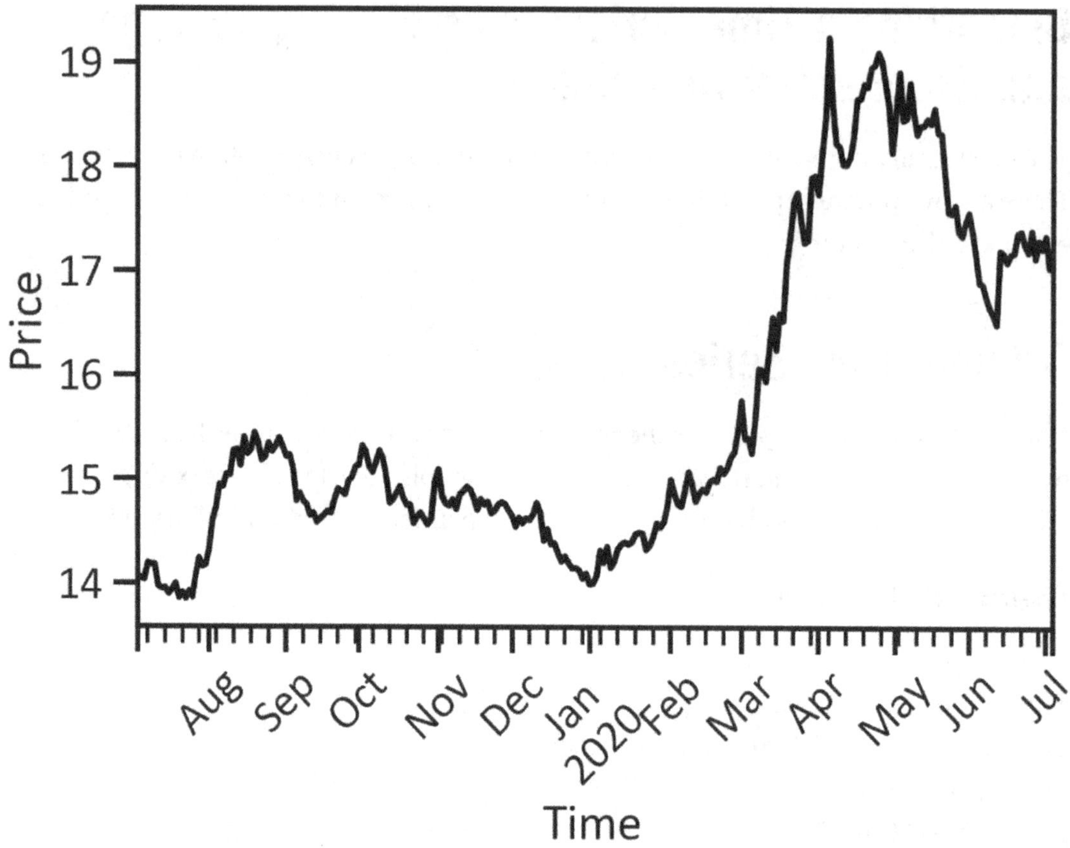

Figure 3-7. *Original time series*

The Moving Averages Smoothing Technique

MA applies the means of weight average of preceding and recent data points. It is useful when there is no seasonality in a series, but we want to forecast future instances. Weighting depends on the stability of a series. If a series is stable, then recent data points weigh high and preceding data points weigh low. In contrast, if the series is not stable, then we weigh recent data points that weigh lower and preceding data points higher. Listing 3-9 returns 90 days of moving averages (see Figure 3-8).

Listing 3-9. Time Series (90 Days of Moving Averages)

```
MA90 = df["Close"].rolling(window=90).mean()
df.plot(kind="line",color="navy")
MA90.plot(kind="line",color="black",label="Rolling MA 90 days")
```

```
plt.xlabel("Date")
plt.xticks(rotation=45)
plt.ylabel("Price")
plt.legend()
plt.show()
```

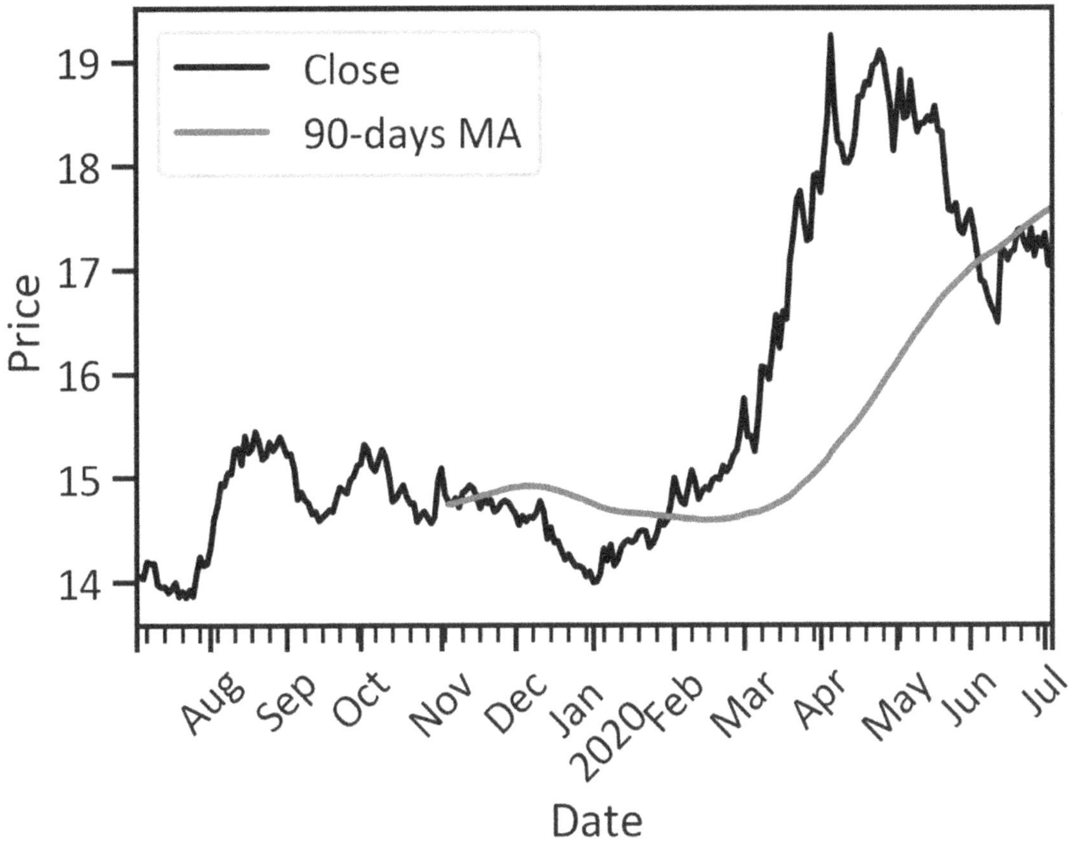

***Figure 3-8.** Time series (90 days of moving averages)*

Figure 3-8 does not consider seasonality. The close price is mostly above the moving average. However, it struggles to keep at abreast with the trend.

The Exponential Smoothing Technique

An alternative to the MA method is exponential smoothing. It weights values outside the window to zero. Using this method, large weighted values rapidly die out, and small weighted values slowly die out. Listing 3-10 specifies the half life (the time lag at which the exponential weights decay by one-half) as 30 (see Figure 3-9).

Listing 3-10. Time Series (Exponential)

```
Exp = df["Close"].ewm(halflife=30).mean()
fig, ax = plt.subplots()
Original.plot(kind="line", color="navy")
Exp.plot(kind="line", color="gray", label="Half Life")
plt.xlabel("Time")
plt.ylabel("Close")
plt.xticks(rotation=45)
plt.legend()
plt.show
```

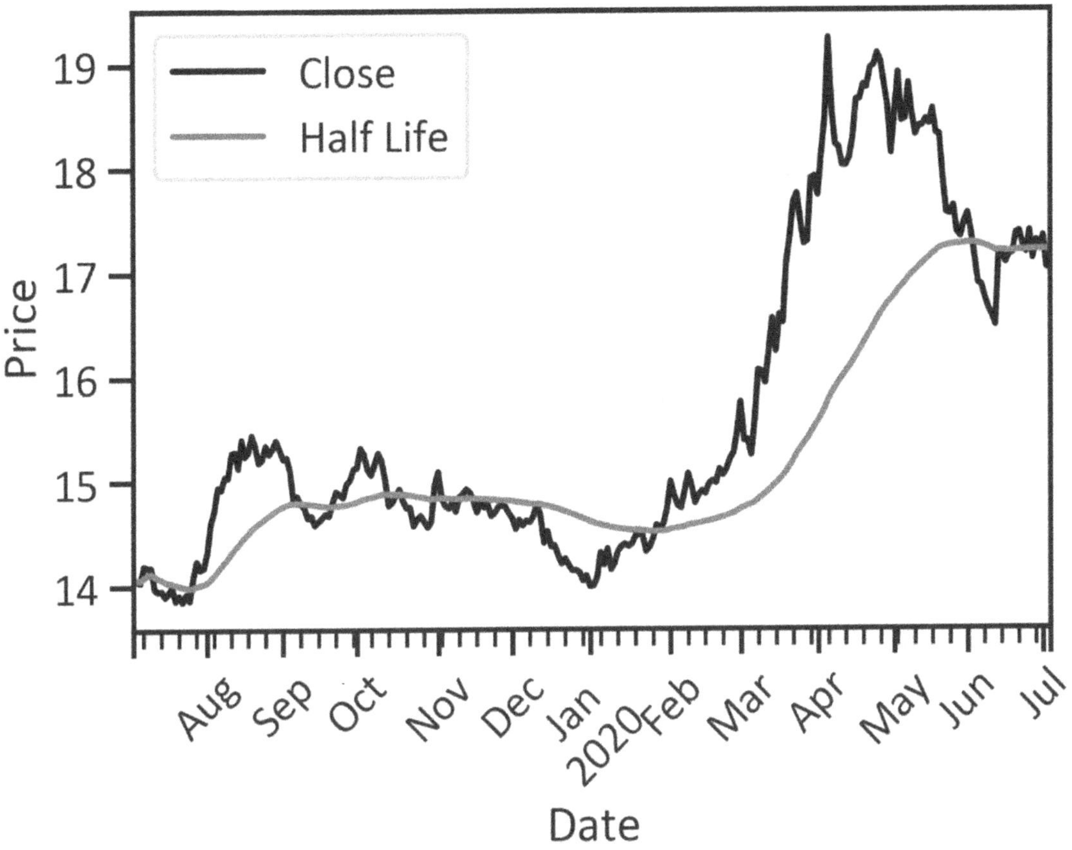

***Figure 3-9.** Time series (exponential)*

Smoothing methods do not produce confidence intervals for forecasts or tests for whether the model is reasonable.

Running Minimum and Maximum Value

Running min is the lowest value in a series as it progresses, and running max is the highest value in a series as it progresses. Extreme values of a series tell us much about boundaries in a series. Listing 3-11 estimates the running min and max using the `expanding()` method (see Figure 3-10).

Listing 3-11. Running Minimum and Maximum Values

```
expanding_df = df
expanding_df["running_min"] = expanding_df["Close"].expanding().min()
expanding_df["running_max"] = expanding_df["Close"].expanding().max()
fig, ax = plt.subplots()
Original.plot(kind="line", color="navy")
expanding_df["running_min"].plot(kind="line", color="red")
expanding_df["running_max"].plot(kind="line", color="red")
plt.xlabel("Date")
plt.ylabel("Close")
plt.legend()
plt.xticks(rotation=45)
plt.show()
```

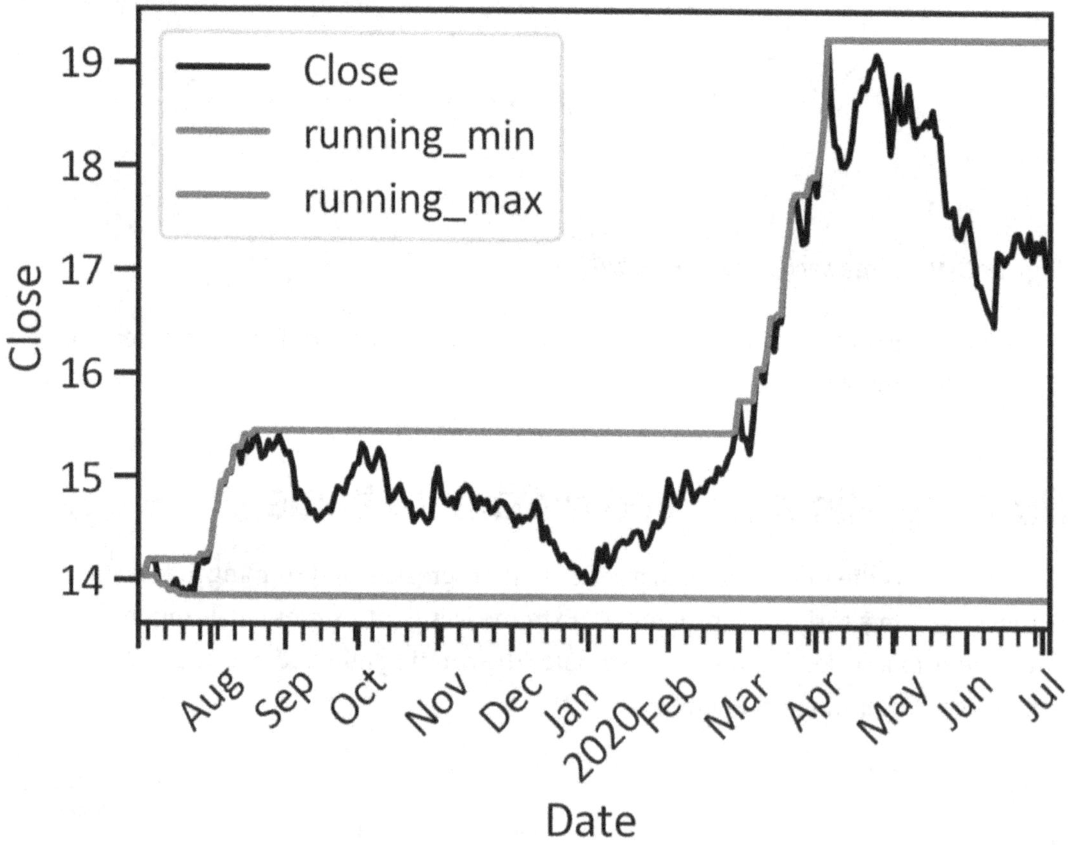

Figure 3-10. *Running the minimum and maximum value of price*

Figure 3-10 shows that the USD/ZAR closing price dropped to a low of 13.8531 on July 24, 2019. Thereafter, it approached a psychological level. In the last quarter of 2019, the South African rand moderately steadied against the US dollar. However, those gains quickly wiped out at the beginning of the year. The US dollar edged higher, reaching a new high of 19.2486 on April 6, 2020.

Find the Rate of Return and Rolling Rate of Return

Returns refer to profit on an investment. It shows the change in the value of an investment. Listing 3-12 shows the delete expansion.

Listing 3-12. Delete Expansion

```
del df_expanding["Running_min"]
del df_expanding["Running_max"]
```

Listing 3-13 and Figure 3-11 show the rate of return of the portfolio.

Listing 3-13. Rate of Return

```
pr = df.pct_change()
pr_plus_one = pr.add(1)
cumulative_return = pr_plus_one.cumprod().sub(1)
fig, ax = plt.subplots()
cumulative_return.mul(100).plot(ax=ax, color="navy")
plt.xlabel("Date")
plt.ylabel("Return (%)")
plt.xticks(rotation=45)
plt.show()
```

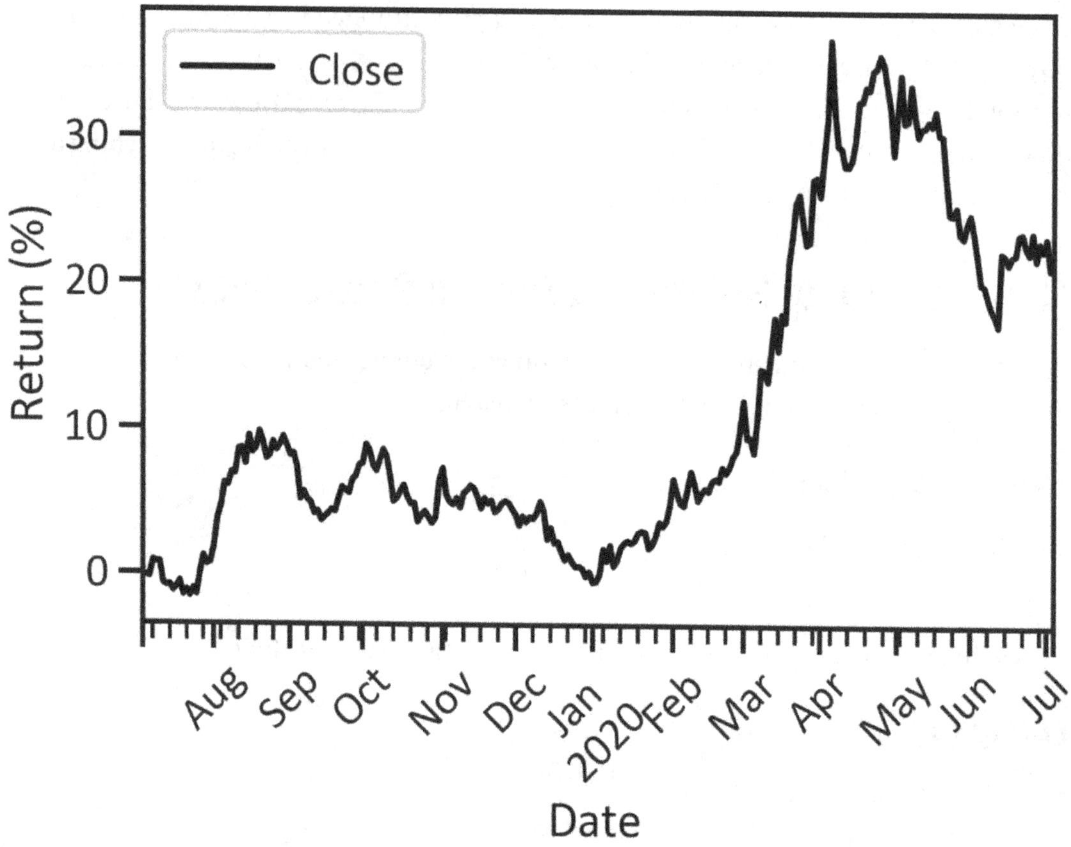

Figure 3-11. *Rate of return*

Find the Rolling Rate of Return

The rolling rate of return shows the rolling window rate of return. It communicates the stability of the rate of return. Listing 3-14 finds a linear combination of the previous mean returns, and Listing 3-15 plots 90 days of the rolling rate of return (see Figure 3-12).

Listing 3-14. Rolling Rate of Return Function

```
def get_period_return(period_return):
    return np.prod(period_return + 1) - 1
```

Listing 3-15. Rolling Rate of Return

```
rolling_90_days_rate_of_return = df["Close"].rolling(window=90).apply(
get_period_return)
fig, ax = plt.subplots()
rolling_90_days_rate_of_return.plot(color="navy")
plt.xlabel("Date")
plt.ylabel("Return (%)")
plt.xticks(rotation=45)
plt.show()
```

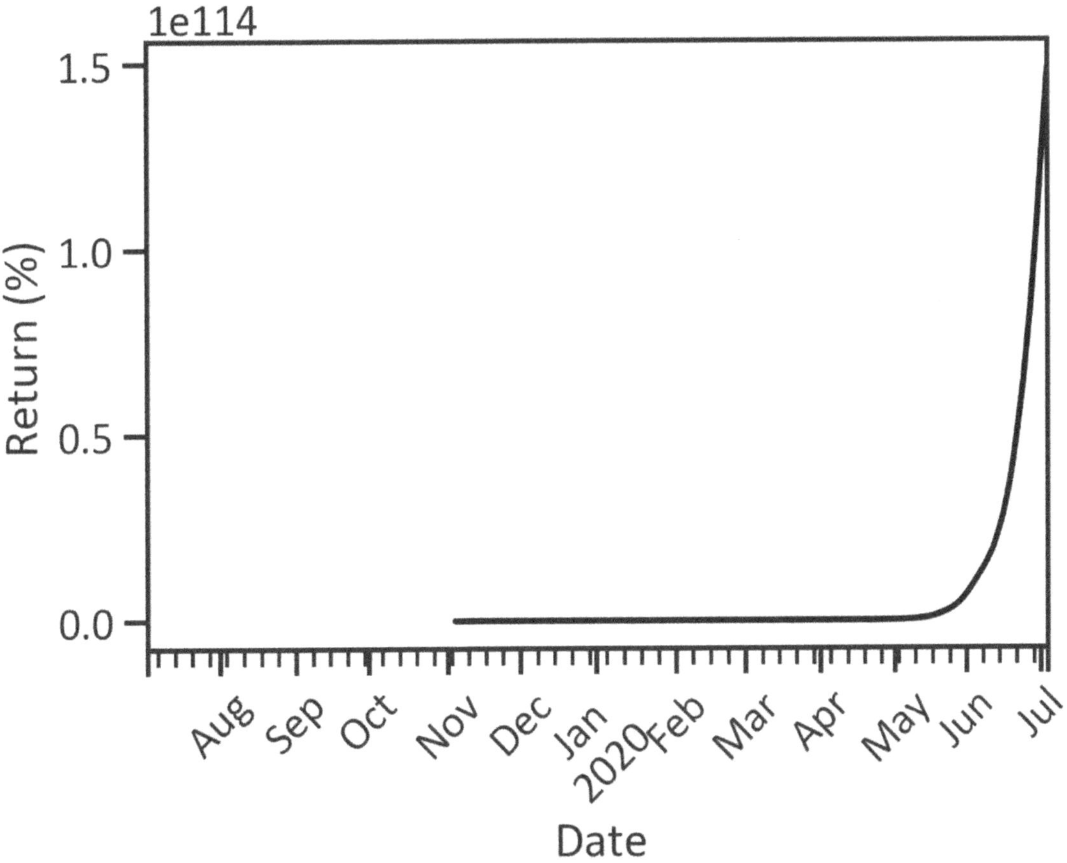

Figure 3-12. *Rolling rate of return*

SARIMAX Hyperparameter Optimization

Listing 3-16 splits the data into training data and test data.

Listing 3-16. Split Data

```
train = df.iloc[:len(df) -12]
test = df.iloc[:len(df) -12]
start = len(train)
end = len(train) + len(train) - 1
```

Listing 3-17 applies the `itertools` package to determine the optimal parameters of the seasonal ARIMA model using the akaike information criterion (AIC).

Listing 3-17. Hyperparameter Optimization

```
import itertools
p = d = q = range(0, 2)
pdq = list(itertools.product(p, d, q))
seasonal_pdq = [(x[0], x[1], x[2], 12) for x in list(itertools.product
(p, d, q))]
for param in pdq:
    for param in pdq:
        for param_seasonal in seasonal_pdq:
            try:
                mod = sm.tsa.statespace.SARIMAX(train["Close"],
                                                order=param, seasonal_
                                                order=param_seasonal,
                                                enforce_stationarity=False,
                                                enforce_
                                                invertibility=False)
                results = mod.fit()
                print('SARIMAX{}x{}12 - AIC:{}'.format(param, param_
                seasonal, results.aic))
            except:
                continue
```

AIC is defined as AIC = -2 lnL + 2k, where lnL is the maximized log-likelihood of the model, and k is the number of parameters estimated. We identified that the optimal parameters for the seasonal ARIMA model are as follows: - ARIMA (1, 1, 1) x(1, 1, 1, 12)12.

Develop the SARIMAX Model

Listing 3-18 configures the ARIMA model with a (1, 1, 1) order and (1, 1, 1, 12) seasonal order.

Listing 3-18. Finalize the SARIMAX Model

```
timeseriesmodel1 = sm.tsa.statespace.SARIMAX(new_df["Close"], order=(1, 1, 1),
                                             seasonal_order=(1, 1, 1, 12),
                                             enforce_stationarity=False,
                                             enforce_invertibility=False)
timeseriesmodel_fit1 = timeseriesmodel1.fit()
```

Model Summary

Listing 3-19 and Table 3-2 show the profile of the SARIMAX model.

Listing 3-19. SARIMAX Profile

```
timeseriesmodel_fit1.summary()
```

Table 3-2. SARIMAX Profile

Dep. Variable:	Close	**No. Observations:**	263
Model:	SARIMAX(1, 1, 1)x(1, 1, 1, 12)	**Log Likelihood**	53.382
Date:	Fri, 16 Oct 2020	**AIC**	-96.765
Time:	16:07:34	**BIC**	-79.446
Sample:	07-03-2019	**HQIC**	-89.783
	- 07-03-2020		
Covariance Type:	Opg		

	coef	**std err**	**Z**	**P>lzl**	**[0.025**	**0.975]**
ar.L1	0.4856	1.971	0.246	0.805	-3.377	4.348
ma.L1	-0.5019	1.952	-0.257	0.797	-4.327	3.324
ar.S.L12	-4.739e-06	0.008	-0.001	1.000	-0.015	0.015
ma.S.L12	-0.9322	0.046	-20.178	0.000	-1.023	-0.842
sigma2	0.0348	0.002	15.107	0.000	0.030	0.039

Ljung-Box (Q):	66.32	**Jarque-Bera (JB):**	91.15
Prob(Q):	0.01	**Prob(JB):**	0.00
Heteroskedasticity (H):	5.28	**Skew:**	0.53
Prob(H) (two-sided):	0.00	**Kurtosis:**	5.86

Table 3-2 shows that the AIC score is -96.765, that the score is low compared to other orders, and that ar.L1, ma.L1, and ar.S.L12 are not statistically significant. The p-value scores are greater than 0.005. This shows that the SARIMAX (1, 1, 1, 12) model is not the appropriate model. It is advisable to try all orders and seasonal orders to find the most hyperparameters that produce a model with significant accurate predictions.

Forecast a SARIMAX

Listing 3-20 produces future instances of the series and plots them with the actual values (see Figure 3-13).

Listing 3-20. Forecast

```
predictions1 = timeseriesmodel_fit1.predict(start,end,typ="levels").
rename("Prediction")
test.plot(color="navy")
predictions1.plot(color="red")
plt.ylabel("Price")
plt.xlabel("Date")
plt.xticks(rotation=45)
plt.show()
```

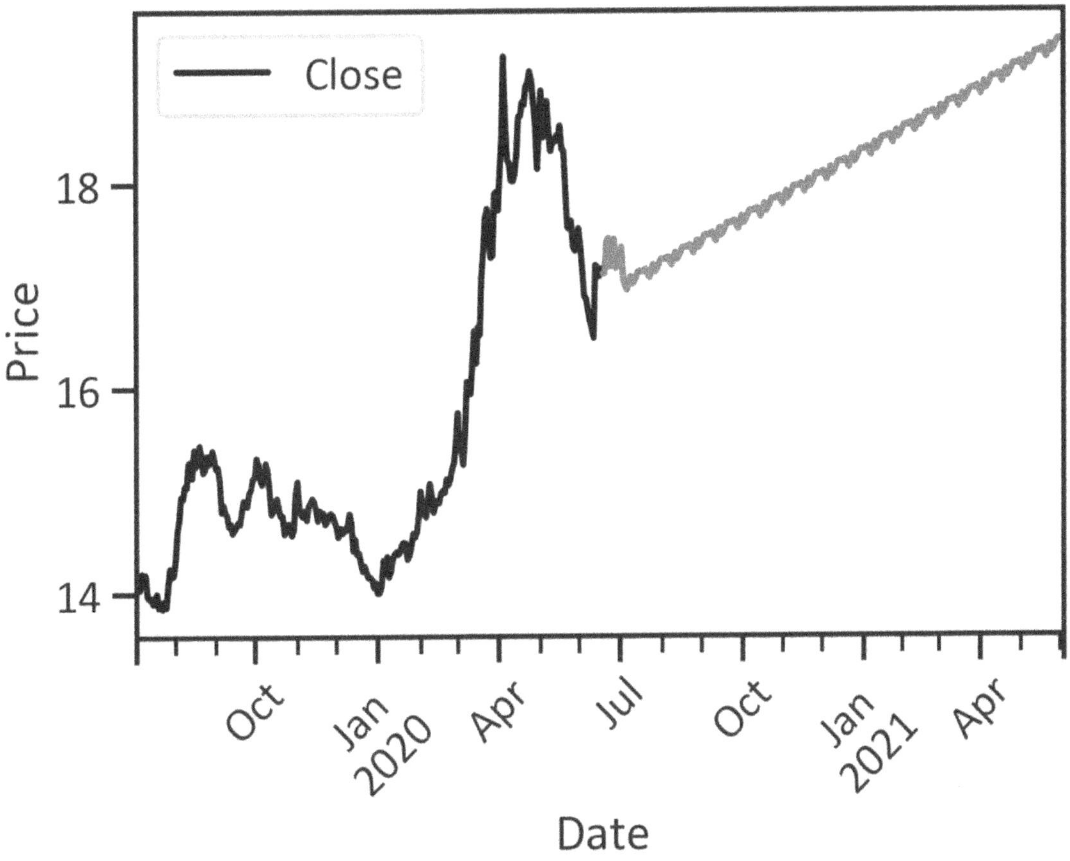

Figure 3-13. *SARMIAX (1, 1, 1, 12) forecast*

Conclusion

This chapter covers the time-series analysis. We trained a SARIMAX model with a (1, 1, 1) order and a (1, 1, 1, 12) seasonal order. It forecasts an upward trend; however, it is unreliable because of influential data points in the series. The model commits significant mistakes when it forecasts the series. The subsequent chapter introduces the additive model.

CHAPTER 4

High-Quality Time-Series Analysis

The preceding chapter covered seasonal ARIMA. After all the considerable effort in data preprocessing and hyperparameter optimization, the model generates considerable errors when forecasting future instances of the series. For a fast and automated forecasting procedure, use Facebook's Prophet; it forecasts time-series data based on nonlinear trends with seasonality and holiday effects. This chapter introduces Prophet and presents a way of developing and testing an additive model. First, it discusses the crucial difference between the Statsmodels package and the Prophet package.

The Difference Between Statsmodels and Prophet

Time-series analysis typically requires missing values and outliers diagnosis and treatment, multiple test statistics to verify key assumptions, hyperparameter optimization, and seasonal effects control. If we commit a slight mistake in the workflow, then the model will make significant mistakes when forecasting future instances. Consequently, building a model using Statsmodels reasonably requires a certain level of mastery.

Fondly remember that machine learning is about inducing a computer with intelligence using minimal code. Statsmodels does not offer us that. Prophet fills that gap. It was developed by Facebook's Core Data Science team. It performs tasks such as missing value and outlier detection, hyperparameter optimization, and control of seasonality and holiday effects. To install FB Prophet in the Python environment, use `pip install fbprophet`. To install it in the Conda environment, use `conda install -c conda-forge fbprophet`.

© Tshepo Chris Nokeri 2021
T. C. Nokeri, *Data Science Revealed*, https://doi.org/10.1007/978-1-4842-6870-4_4

Understand Prophet Components

In the previous chapter, we painstakingly built a time-series model with three components, namely, trend, seasonality, and irregular components.

Prophet Components

The Prophet package sufficiently takes into account the trend, seasonality and holidays, and events. In this section, we will briefly discuss the key components of the package.

Trend

A trend is a single and consistent directional movement (upward or downward). We fit a time-series model to discover a trend. Table 4-1 highlights tunable trend parameters.

Table 4-1. *Tunable Trend Parameters*

Parameters	Description
`growth`	Specifies a piecewise linear or logistic trend
`n_changepoints`	The number of changes to be automatically included, if changepoints are not specified
`change_prior_scale`	The change of automatic changepoint selection

Changepoints

A changepoint is a penalty term. It alters the behavior of a time-series model. If you do not specify changepoints, the default changepoint value is automated.

Seasonality

Seasonality represents consistent year-to-year upward or downward movements. Equation 4-1 approximates this.

$$S(t) = \sum_{n=1}^{N} \alpha_n \, cos \, cos\left(\frac{2\pi nt}{P}\right) + b_n \cos\left(\frac{2\pi nt}{P}\right) \qquad \text{(Equation 4-1)}$$

Here, *P* is the period (7 for weekly data, 30 for monthly data, 90 for quarterly data, and 365 for yearly data).

Holidays and Events Parameters

Holidays and events affect the time series' conditional mean. Table 4-2 outlines key tunable holiday and event parameters.

Table 4-2. *Holiday and Event Parameters*

Parameters	Description
`daily_seasonality`	Fit daily seasonality
`weekly_seasonality`	Fit weekly seasonality
`year_seasonality`	Fit yearly seasonality
`holidays`	Include holiday name and date
`yeasonality_prior_scale`	Determine strength of needed for seasonal or holiday components

The Additive Model

The additive models assume that the trend and cycle are treated as one term. Its components are similar to the Holt-Winters technique. We express the equation as shown in Equation 4-2.

$$y = S(t) + T(t) * I(t) \qquad \text{(Equation 4-2)}$$

The formula is written mathematically as shown in Equation 4-3.

$$y = g(t) + s(t) + h(t) + \varepsilon_i \qquad \text{(Equation 4-3)}$$

Here, *g(t)* represents the linear or logistic growth curve for modeling changes that are not periodic, *s(t)* represents the periodic changes (daily, weekly, yearly seasonality), *h(t)* represents the effects of holidays, and $+ \varepsilon_i$ represents the error term that considers unusual changes.

Data Preprocessing

We obtained the example data from Yahoo Finance.[1] Before training the model, repurpose the data as follows:

- `df["ds"]`, which repurposes time
- `df["y"]`, which repurposes the independent variable
- `df.set_index("")`, which sets the date and time as the index column

Listing 4-1 sets column names to the right format.

Listing 4-1. Process Data

```
df["ds"] = df["Date"]
df["y"] = df["Close"]
df.set_index("Date")
```

Table 4-3 shows how a dataframe should look like before developing a time-series model using Prophet.

Table 4-3. *Dataset*

	Close	ds	y
Date			
2019-07-03	14.074300	2019-07-03	14.074300
2019-07-04	14.052900	2019-07-04	14.052900
2019-07-05	14.038500	2019-07-05	14.038500
2019-07-08	14.195200	2019-07-08	14.195200
2019-07-09	14.179500	2019-07-09	14.179500
...	...	...	...
2020-06-29	17.298901	2020-06-29	17.298901
2020-06-30	17.219200	2020-06-30	17.219200

(*continued*)

[1] `https://finance.yahoo.com/quote/usdzar=x/`

Table 4-3. *(continued)*

	Close	ds	y
2020-07-01	17.341900	2020-07-01	17.341900
2020-07-02	17.039301	2020-07-02	17.039301
2020-07-03	17.037100	2020-07-03	17.037100

Develop the Prophet Model

Listing 4-2 specifies the official holidays. Listing 4-3 configures the model with a confidence interval to 95%, and it considers yearly seasonality, weekly seasonality, and daily seasonality.

Listing 4-2. Specify Holidays

```
holidays = pd.DataFrame({
  'holiday': 'playoff',
  'ds': pd.to_datetime(["2020-12-25", "2020-12-24", "2020-12-23",
  "2019-12-25", "2021-01-01", "2021-01-20"]),
    "lower_window": 0,
    "upper_window": 1,
})
```

Listing 4-3 completes the model.

Listing 4-3. Develop Prophet Model

```
m = Prophet(holidays=holidays,
            interval_width=0.95,
            yearly_seasonality=True,
            weekly_seasonality=True,
            daily_seasonality=True,
            changepoint_prior_scale=0.095)
m.add_country_holidays(country_name='US')
m.fit(df)
```

Create the Future Data Frame

Listing 4-4 applies the make_future_dataframe() method to create the future dataframe.

Listing 4-4. Create a Future Data Frame Constrained to 365 Days

```
future = m.make_future_dataframe(periods=365)
```

Forecast

Listing 4-5 applies the predict() method to forecast future instances.

Listing 4-5. Forecast Time Series

```
forecast = m.predict(future)
```

Listing 4-6 plots previous values and forecasted values (see Figure 4-1).

Listing 4-6. Forecast

```
m.plot(forecast)
plt.xlabel("Date")
plt.ylabel("Price")
plt.xticks(rotation=45)
plt.show()
```

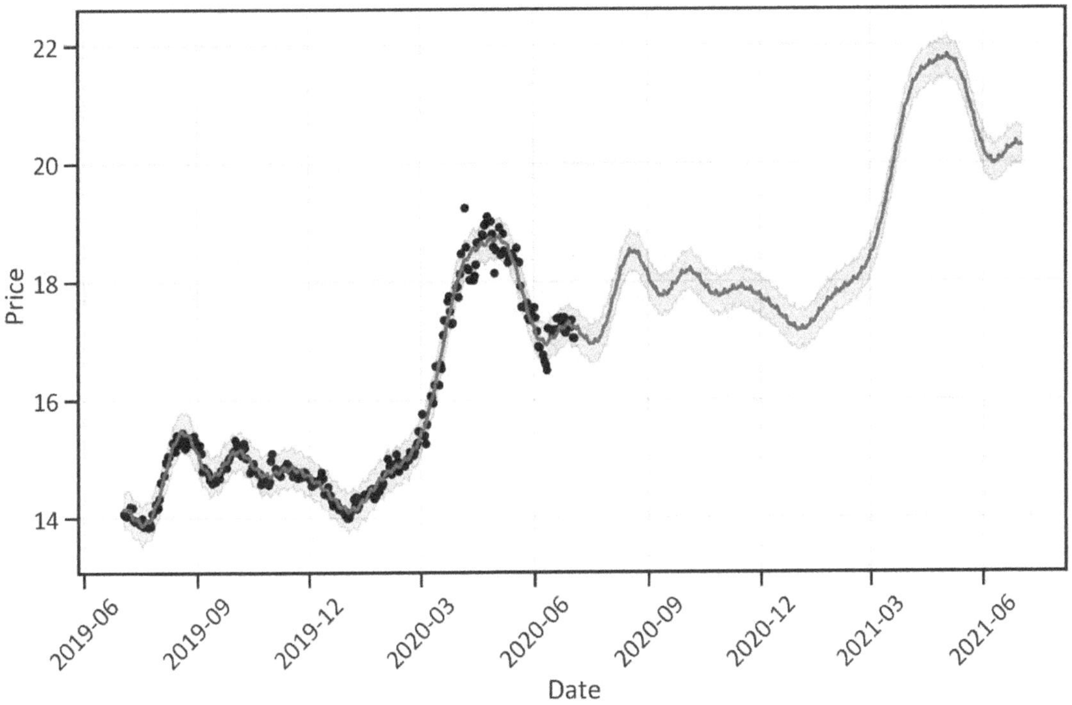

Figure 4-1. *Forecast*

Figure 4-1 tacitly agrees with the SARIMAX in the previous chapter; however, it provides more details. It forecasts a long-run bullish trend in the first quarters of the year 2021.

Seasonal Components

Listing 4-7 decomposes the series (see Figure 4-2).

Listing 4-7. Seasonal Components

```
model.plot_components(forecast)
plt.show()
```

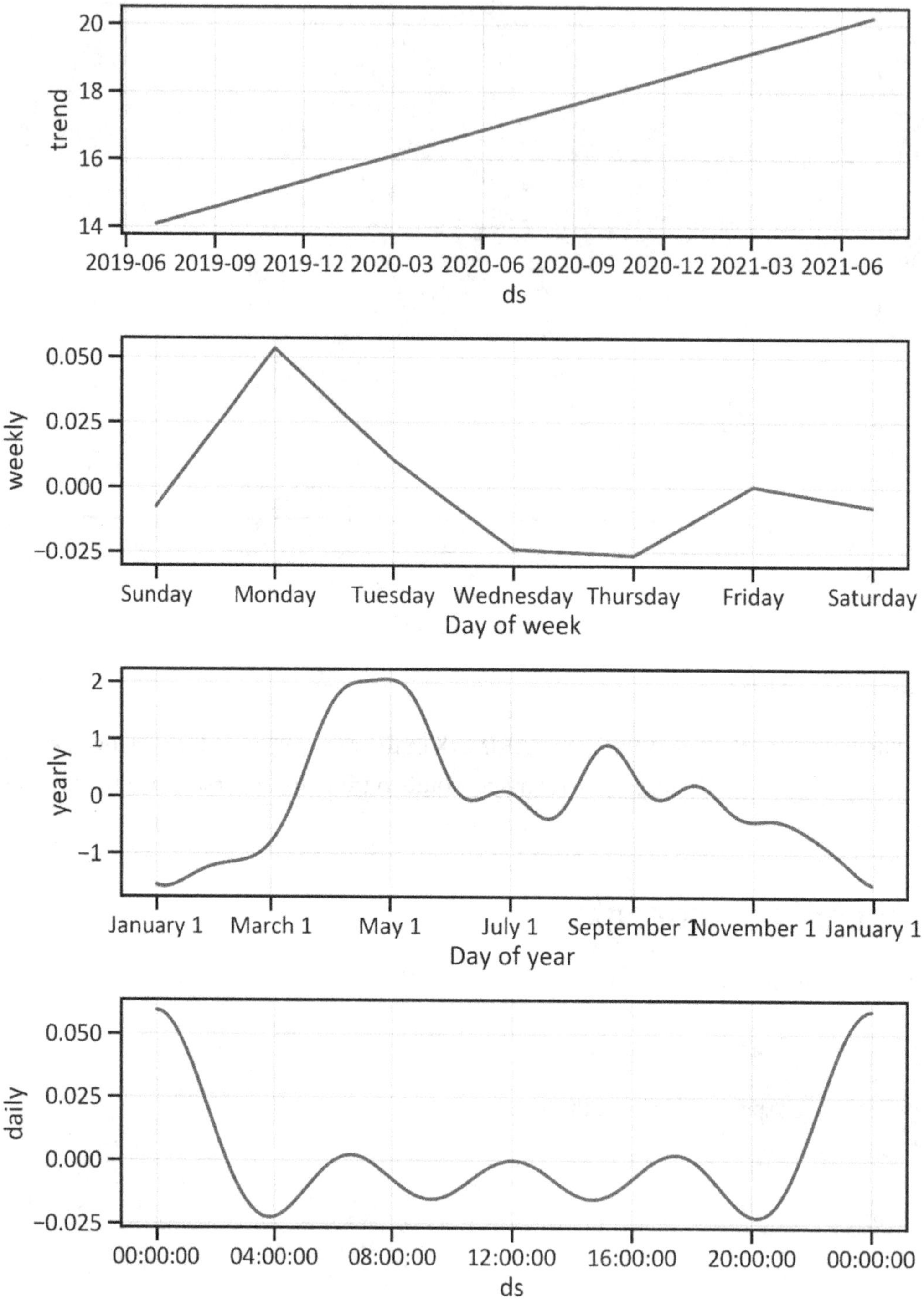

Figure 4-2. *Seasonal components*

Cross-Validate the Model

Cross validation tests model performance. In the background, Prophet finds forecast errors of historical data. It sets the cutoff point and only data up to that specific cutoff point. Listing 4-8 applies the `cross_validation()` method with a specified initial starting point of the training data (initial), forecast horizon (horizon), and space between cutting off points (period). See Table 4-4.

Listing 4-8. Cross Validation

```
from fbprophet.diagnostics import cross_validation
df_cv = cross_validation(model, initial="210 days",period="15 days",
horizon="70 days") df_cv
```

Table 4-4. *Cross-Validation Table*

	ds	yhat	yhat_lower	yhat_upper	y	cutoff
0	2020-02-10	14.706377	14.527866	14.862567	15.071000	2020-02-09
1	2020-02-11	14.594084	14.429592	14.753000	14.953500	2020-02-09
2	2020-02-12	14.448900	14.283988	14.625561	14.791600	2020-02-09
3	2020-02-13	14.258331	14.094124	14.428341	14.865500	2020-02-09
4	2020-02-14	14.028495	13.858400	14.204660	14.905000	2020-02-09
...	...	...	...	...	...	...
295	2020-06-29	19.919252	19.620802	20.193089	17.298901	2020-04-24
296	2020-06-30	19.951939	19.660207	20.233733	17.219200	2020-04-24
297	2020-07-01	19.966822	19.665303	20.250541	17.341900	2020-04-24
298	2020-07-02	20.012227	19.725297	20.301380	17.039301	2020-04-24
299	2020-07-03	20.049481	19.752089	20.347799	17.037100	2020-04-24

Evaluate the Model

Listing 4-9 applies the `performance_metrics()` method to return key evaluation metrics (see Table 4-5).

Listing 4-9. Performance

```
from fbprophet.diagnostics import performance_metrics
df_p = performance_metrics(df_cv)
df_p
```

Table 4-5. *Performance Metrics*

	horizon	mse	Rmse	mae	mape	mdape	coverage
0	7 days	0.682286	0.826006	0.672591	0.038865	0.033701	0.166667
1	8 days	1.145452	1.070258	0.888658	0.051701	0.047487	0.100000
2	9 days	1.557723	1.248088	1.077183	0.063369	0.056721	0.033333
3	10 days	2.141915	1.463528	1.299301	0.077282	0.066508	0.033333
4	11 days	3.547450	1.883468	1.648495	0.097855	0.086243	0.000000
...	...	...	...	...	...	...	...
59	66 days	187.260756	13.684325	9.480800	0.549271	0.244816	0.000000
60	67 days	157.856975	12.564115	8.915581	0.515198	0.244816	0.000000
61	68 days	137.029889	11.705977	8.692623	0.499211	0.253436	0.000000
62	69 days	116.105651	10.775233	8.146737	0.466255	0.252308	0.000000
63	70 days	96.738282	9.835562	7.483452	0.427025	0.227503	0.000000

Conclusion

This chapter covers the generalized additive model. The model takes seasonality into account and uses time as a regressor. Its performance surpasses that of the seasonal ARIMA model. The model commits minor errors when forecasting future instances of the series. We can rely on the Prophet model to forecast a time series.

The first four chapters of this book properly introduce the parametric method. This method makes bold assumptions about the underlying structure of the data. It assumes the underlying structure of the data is linear.

The subsequent chapter introduces the nonparametric method. This method supports flexible assumptions about the underlying structure of the data. It assumes the underlying structure of the data is nonlinear.

CHAPTER 5

Logistic Regression Analysis

This chapter covers the logistic regression concept and implementation in a structured way. Preceding chapters introduced supervised learning and concentrated on the parametric method. In supervised learning, we present a model with a set of correct answers, and we then allow a model to predict unseen data. We use the parametric method to solve regression problems (when a dependent variable is a continuous variable).

This chapter introduces a supervised learning method recognized as the nonparametric method (or the classification method). Unlike the linear regression method, it has flexible assumptions about the structure of the data. It does not have linearity and normality assumptions. There are two intact families of the classification method. They are binary classification, which is used when a dependent variable has two outcomes (i.e., yes/no, pass/fail), and multiclass classification, which is used when a dependent variable has more than two outcomes (negative/neutral/positive).

What Is Logistic Regression?

Logistic regression is a model that estimates the extent to which an independent variable influences a dependent variable. An independent variable is a continuous variable or categorical, and the dependent variable is invariably a categorical variable with only two outcomes. Although logistic regression is called *regression*, it is not a regression model but a classification model. Like with linear regression, we use the model to find an intercept and slope.

© Tshepo Chris Nokeri 2021
T. C. Nokeri, *Data Science Revealed*, https://doi.org/10.1007/978-1-4842-6870-4_5

Logistic Regression Assumptions

The classification method does not contain strict assumptions about the structure of the data. It assumes that a dependent variable is a categorical variable and there are more than 50 data points in the data. Ideally, we require a large sample.

Logistic Regression in Practice

We obtained the example data from Kaggle.[1] Table 5-1 shows the first rows of the data.

Table 5-1. *Dataset*

	Pregnancies	Glucose	Blood Pressure	Skin Thickness	Insulin	BMI	Diabetes Pedigree Function	Age	Outcome
0	6	148	72	35	0	33.6	0.627	50	1
1	1	85	66	29	0	26.6	0.351	31	0
2	8	183	64	0	0	23.3	0.672	32	1
3	1	89	66	23	94	28.1	0.167	21	0
4	0	137	40	35	168	43.1	2.288	33	1

Sigmoid Function

Logistic regression uses the sigmoid function to transform an output in such a way that it promptly returns a probabilistic value that is assigned to two categorical classes. The standard formula of is an S-shape or sigmoid shape. Figure 5-1 shows a standard logistic function.

[1] https://www.kaggle.com/uciml/pima-indians-diabetes-database

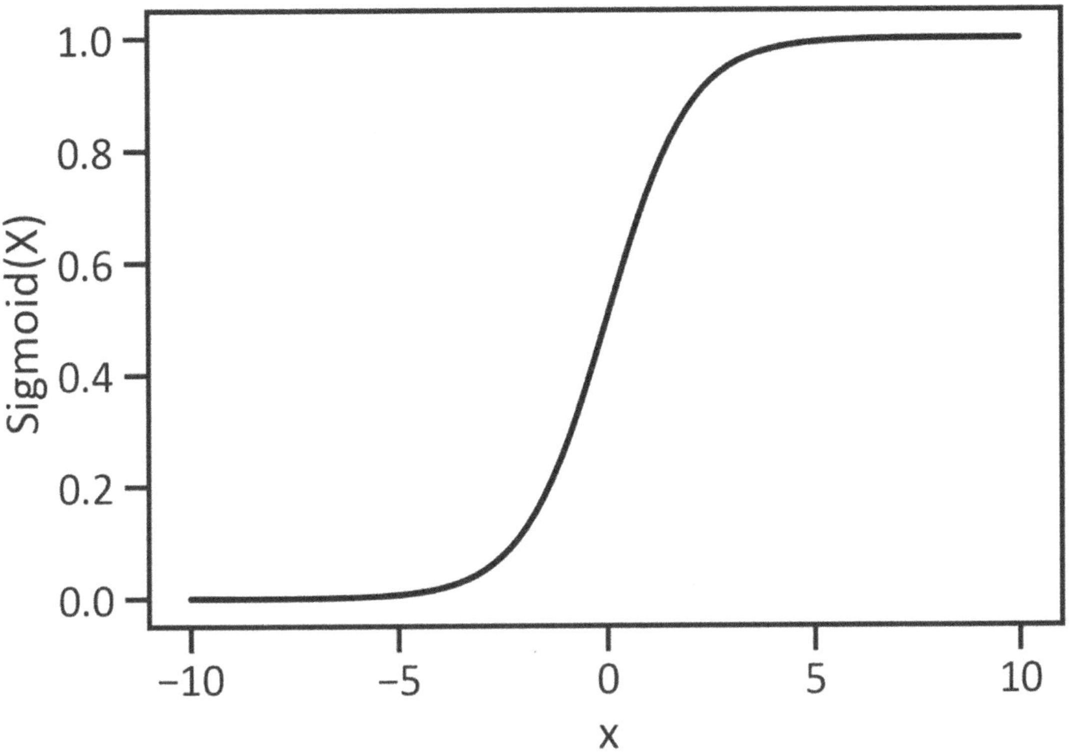

Figure 5-1. *Sigmoid function*

Equation 5-1 expresses the sigmoid function.

$$S(x) = \frac{L}{1 - e^{-x}} = \frac{e^x}{e^x + 1} \qquad \text{(Equation 5-1)}$$

The standard sigmoid function takes any input and outputs a value between 0 and 1.

Descriptive Analysis

Listing 5-1 produces descriptive statistics (see Table 5-2).

Listing 5-1. Descriptive Statistics

```
df.describe().transpose()
```

Table 5-2. *Descriptive Statistics*

	count	mean	std	min	25%	50%	75%	Max
Pregnancies	768.0	3.845052	3.369578	0.000	1.00000	3.0000	6.00000	17.00
Glucose	768.0	120.894531	31.972618	0.000	99.00000	117.0000	140.25000	199.00
BloodPressure	768.0	69.105469	19.355807	0.000	62.00000	72.0000	80.00000	122.00
SkinThickness	768.0	20.536458	15.952218	0.000	0.00000	23.0000	32.00000	99.00
Insulin	768.0	79.799479	115.244002	0.000	0.00000	30.5000	127.25000	846.00
BMI	768.0	31.992578	7.884160	0.000	27.30000	32.0000	36.60000	67.10
DiabetesPedigreeFunction	768.0	0.471876	0.331329	0.078	0.24375	0.3725	0.62625	2.42
Age	768.0	33.240885	11.760232	21.000	24.00000	29.0000	41.00000	81.00
Outcome	768.0	0.348958	0.476951	0.000	0.00000	0.0000	1.00000	1.00

The Pearson Correlation Method

Listing 5-2 shows off the strength of the linear relationships (see Figure 5-2).

Listing 5-2. Pearson Correlation Matrix

```
dfcorr = df.corr(method="pearson")
sns.heatmap(dfcorr, annot=True,annot_kws={"size":12}, cmap="Blues")
plt.show()
```

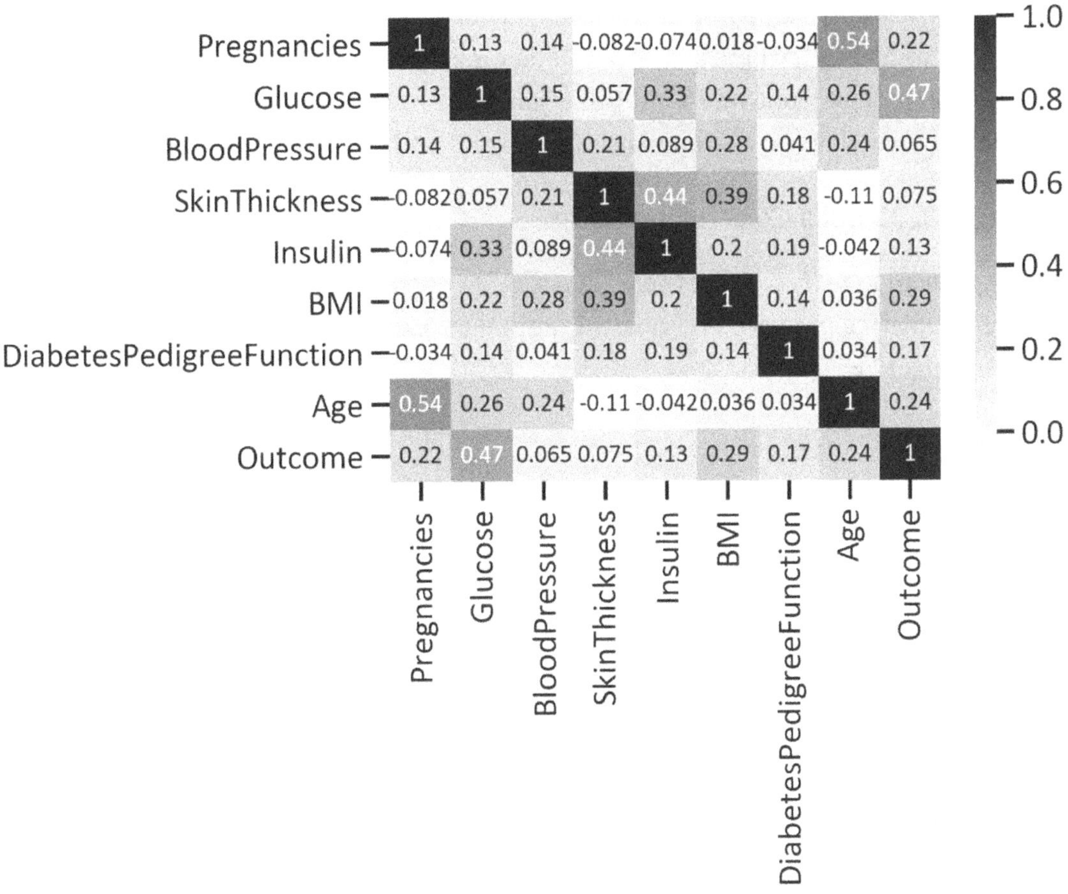

Figure 5-2. *Pearson correlation matrix*

Figure 5-2 shows a weak positive association between most variables in the data.

Other Correlation Methods

There are other methods used to measure correlation between variables, such as the Kendall method and the Spearman method. Their use depends on the use case.

The Kendall Correlation Method

The Kendall correlation method estimates the association between rankings in ordinal variables. An ordinal variable comprises categorical variables with unknown distances. An example of ordinal data is a Likert scale. The Kendall correlation method has values that range from -1 to 1, where -1 indicates that rankings are not the same, 0 indicates that rankings are independent, and 1 indicates that two rankings are the same. Listing 5-3 produces the Kendall correlation matrix (see Figure 5-3).

Listing 5-3. Kendall Correlation Matrix

```
dfcorr_kendall = df.corr(method="kendall")
sns.heatmap(dfcorr_kendall, annot=True,annot_kws={"size":12},
cmap="Oranges")
plt.show()
```

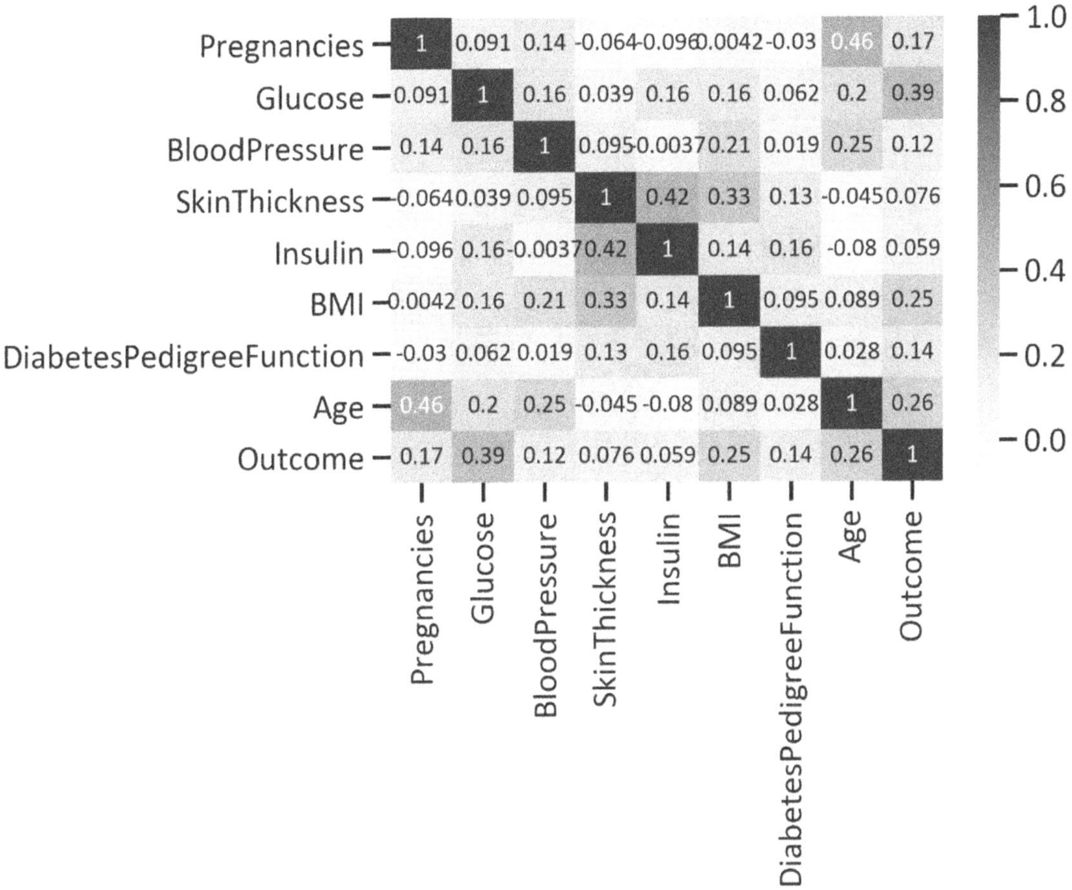

Figure 5-3. *Kendall correlation matrix*

Figure 5-3 shows most rankings are close to being the same. There is a slight difference in the values produced by the Kendall correlation method and the Pearson correlation method.

The Spearman Correlation Method

The Spearman correlation method is an alternative rank correlation method. It measures the statistical dependence between the rankings of two variables. The method helps identify whether variables are linear. It also has values that range from -1 and 1, where -1 indicates that there is a weak association, 0 indicates that there is no association, and 1 indicates that there is a strong association. Listing 5-4 returns a Spearman correlation matrix (see Figure 5-4).

Listing 5-4. Spearman Correlation Matrix

-

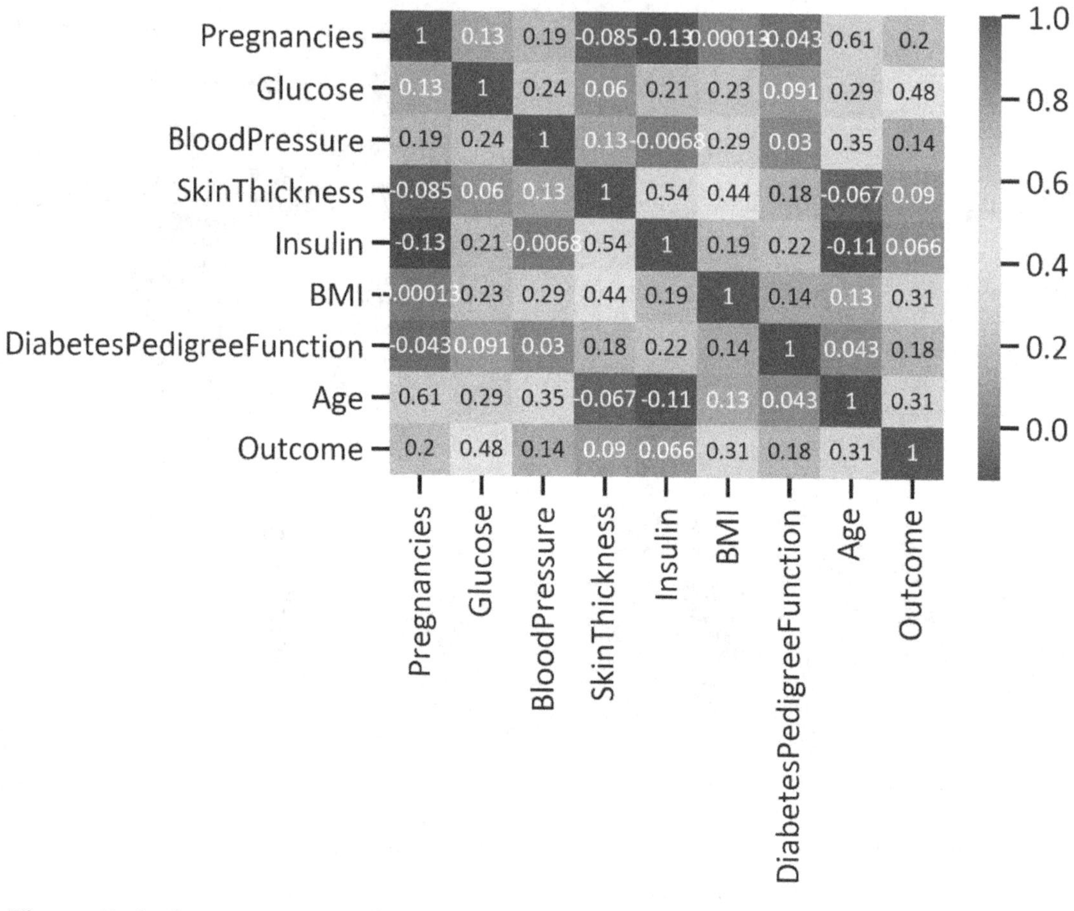

Figure 5-4. *Spearman correlation matrix*

Figure 5-4 also confirms that the underlying structure of the data is linear.

The Covariance Matrix

Covariance gives details about the joint variability between an independent variable and a dependent variable. At most, we are uninterested in the covariance; instead, we are interested in the correlation between variables. Listing 5-5 produces the covariance matrix (see Figure 5-5).

Listing 5-5. Covariance Matrix

```
dfcov = df.cov() sns.heatmap(dfcov, annot=True,annot_kws={"size":12},
cmap="Blues")
plt.show
```

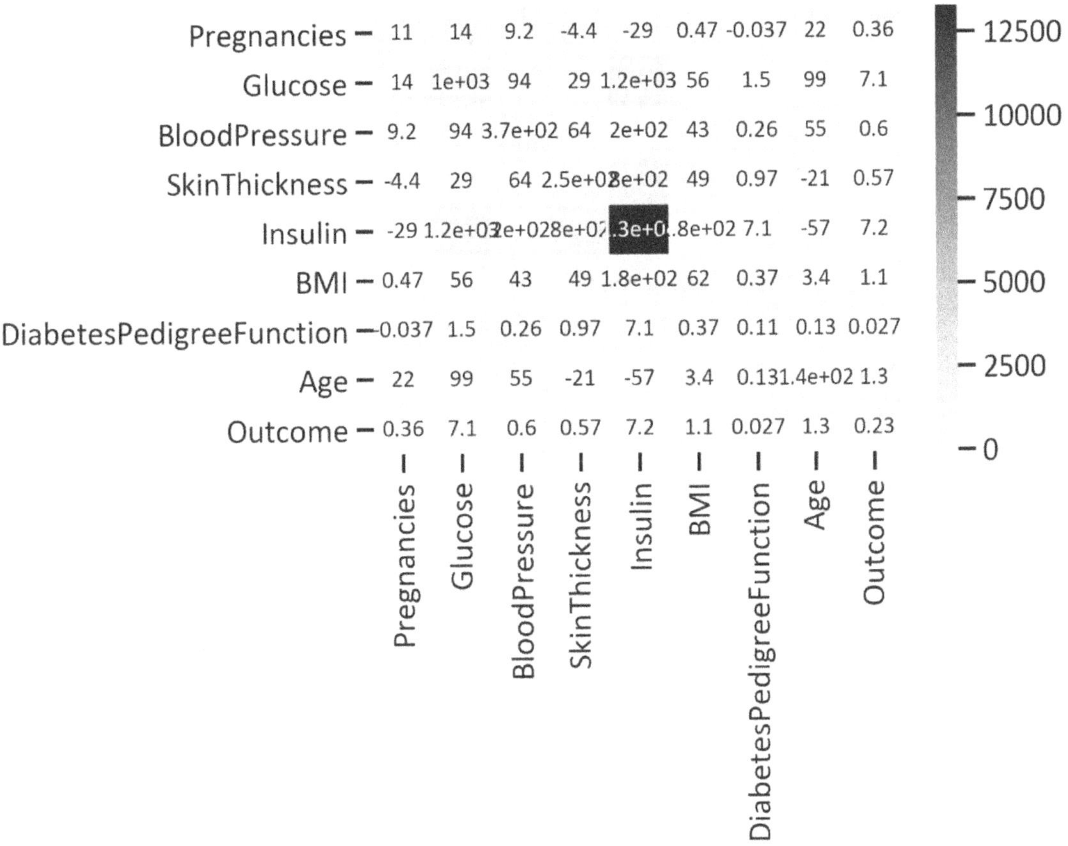

Figure 5-5. *Covariance matrix*

Figure 5-5 depicts variability between variables.

Create X and Y Arrays

After performing descriptive analysis and dimension reduction, the subsequent step is to break down the data into two groups of one-dimensional arrays (x,y), where x represents independent variables and y represents a dependent variable. Listing 5-6 slices the data in such a way that the first eight columns of the data rightfully belong to the x array and the last column belongs to the y array.

Listing 5-6. Create X and Y arrays

```
x = df.iloc[::,0:8]
y = df.iloc[::,-1]
```

Split Data into Training Data and Test Data

In supervised learning models, we split the data into training data and test data. Listing 5-7 splits the data into training data and test data.

Listing 5-7. Split Data

```
from sklearn.model_selection import train_test_split
x_train, x_test, y_train, y_test = train_test_split(x,y, test_size=0.2,
random_state=0)
```

The Eigen Matrix

The most common multicollinearity detection method is the tolerance and variance inflation factor (VIF). Tolerance measures unique total standard variance, and VIF is the model's overall variance including on a single independent variable. A more credible way of finding severity is using eigenvalues. Eigenvalues less than 0 indicate multicollinearity. Eigenvalues between 10 and 100 indicate that there is slight multicollinearity. Last, eigenvalues above 100 show severe multicollinearity. Listing 5-8 generates eigenvectors and values using the NumPy package and passes them as dataframes using Pandas so you can have a clear picture of the matrix (see Table 5-3).

Listing 5-8. Create the Eigen Matrix

```
eigenvalues, eigenvectors = np.linalg.eig(dfcov)
eigenvalues = pd.DataFrame(eigenvalues)
eigenvectors = pd.DataFrame(eigenvectors)
eigens = pd.concat([eigenvalues,eigenvectors],axis=1)
eigens.index = df.columns
eigens.columns = ("Eigen values","Pregnancies","Glucose","BloodPressure",
"SkinThickness","Insulin","BMI","DiabetesPedigreeFunction","Age","Outcome")
eigens
```

Table 5-3. *Eigen Matrix*

	Eigen values	Pregnancies	Glucose	Blood Pressure	Skin Thickness	Insulin	BMI	Diabetes Pedigree Function	Age	Outcome
Pregnancies	13456.577591	-0.002022	0.022650	-0.022465	-0.049041	-0.151620	-0.005077	-0.986468	0.018049	0.010692
Glucose	932.805966	0.097812	0.972186	0.143425	0.119804	0.087995	0.050824	-0.000769	0.006061	0.000643
BloodPressure	390.577869	0.016093	0.141901	-0.922468	-0.262749	0.232160	0.075596	0.001177	-0.002469	-0.000071
SkinThickness	198.184055	0.060757	-0.057856	-0.307012	0.884371	-0.259927	0.221412	0.000379	0.001118	-0.002349
Insulin	112.693345	0.993111	-0.094629	0.020977	-0.065549	0.000169	-0.006137	-0.001426	-0.000072	-0.000295
BMI	45.837383	0.014011	0.046977	-0.132444	0.192811	-0.021518	-0.970677	0.003045	0.013921	0.000672
Diabetes PedigreeFunction	7.763548	0.000537	0.000817	-0.000640	0.002699	-0.001642	-0.002032	0.006306	-0.239281	0.970922
Age	0.164361	-0.003565	0.140168	-0.125454	-0.301007	-0.920492	-0.014979	0.162649	0.003234	-0.001209
Outcome	0.099139	0.000585	0.007010	0.000309	0.002625	-0.006131	-0.013184	-0.019319	-0.970655	-0.239141

Normalize Data

Listing 5-9 transforms the data in such a way that the mean value is 0 and the standard deviation is 1.

Listing 5-9. Normalize Data

```
from sklearn.preprocessing import StandardScaler
scaler = StandardScaler()
x_train = scaler.fit_transform(x_train)
x_test = scaler.transform(x_test)
```

Modeling Data

We develop two models to address the classification problem at hand; the first model uses Statsmodels, and the second one uses SciKit-Learn.

Develop the Logistic Classifier Using Statsmodels

In the past two chapters (covering time-series analysis), we criticized how difficult it is to develop a reliable time-series model using Statsmodels. This does not mean that the package in its entirety is bad. It is still relevant; actually, there are few packages that offer in-depth analysis like Statsmodels does.

Add the Constant

By default, Statsmodels does not include the constant. Listing 5-10 manually adds a constant.

Listing 5-10. Add the Constant

```
from statsmodels.api import sm
x_constant = sm.add_constant(x_train)
x_test = sm.add_constant(x_test)
```

Develop the Logistic Classifier

The default classifier Statsmodels uses the logit model and maximum likelihood method (MLE). We express the formula as shown in Equation 5-2.

$$F_{ML} = log|\Sigma(\theta)| - log|S| + tr\left[S\Sigma(\theta)^{-1}\right] - p \qquad \text{(Equation 5-2)}$$

Here, *log* represents natural logarithms, *S* represents the empirical covariance matrix, Θ represents parameter vectors, and $\sum(\Theta)$ and $|\sum(\Theta)|$ represent the covariance. See Listing 5-11.

Listing 5-11. Develop the Logistic Classifier

```
model = sm.Logit(y_train,x_constant).fit()
```

Partial Regression

The relaxed way of depicting the relationship among variables is using a partial regression plot. Listing 5-12 plots the association between variables under investigation (see Figure 5-6).

Listing 5-12. Partial Regression

```
fig = sm.graphics.plot_partregress_grid(model)
plt.show()
```

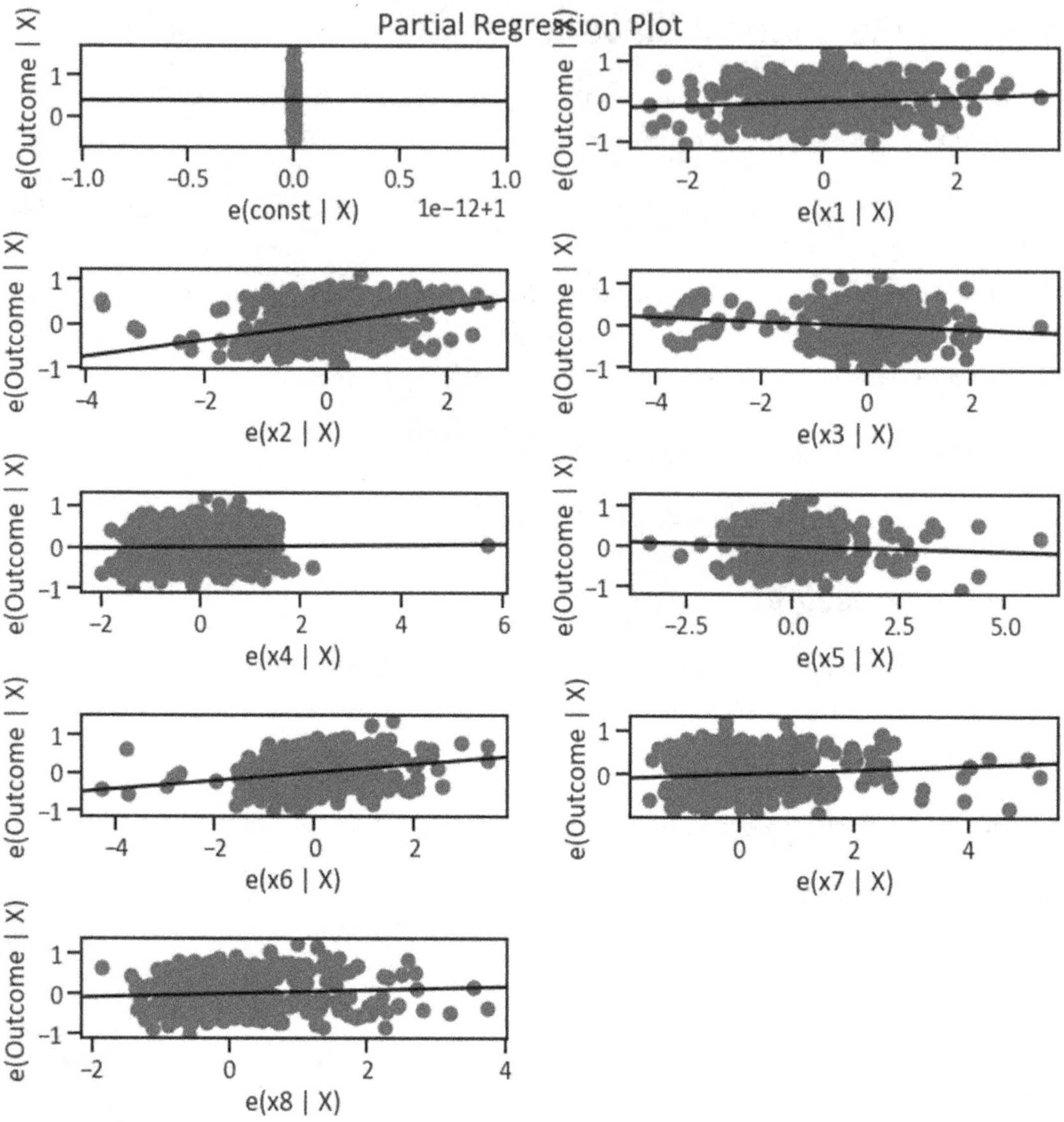

Figure 5-6. *Partial regression grid*

Model Summary

Listing 5-13 and Table 5-4 summarize the model.

Listing 5-13. Profile

```
summary = model.summary()
summary
```

Table 5-4. *Profile*

Dep. Variable:	Outcome	**No. Observations:**	614
Model:	Logit	**Df Residuals:**	605
Method:	MLE	**Df Model:**	8
Date:	Thu, 15 Oct 2020	**Pseudo R-squ.:**	0.2607
Time:	19:15:31	**Log-Likelihood:**	-296.59
converged:	True	**LL-Null:**	-401.18
Covariance Type:	nonrobust	**LLR p-value:**	7.418e-41

	coef	**std err**	**z**	**P>|z|**	**[0.025**	**0.975]**
const	-0.7920	0.106	-7.484	0.000	-0.999	-0.585
x1	0.3147	0.118	2.664	0.008	0.083	0.546
x2	1.0788	0.131	8.224	0.000	0.822	1.336
x3	-0.2685	0.113	-2.369	0.018	-0.491	-0.046
x4	0.0710	0.125	0.569	0.569	-0.174	0.316
x5	-0.1664	0.115	-1.443	0.149	-0.393	0.060
x6	0.6955	0.131	5.313	0.000	0.439	0.952
x7	0.2983	0.110	2.700	0.007	0.082	0.515
x8	0.2394	0.124	1.931	0.054	-0.004	0.483

Table 5-4 displays information about the logistic regression model. It outlines the estimated intercept and slope. In addition, it has test statistics results that are useful for testing hypotheses and information about the model's performance. In classification analysis, we do not use R^2 to measure variability. Instead, we use the pseudo R^2. Table 5-4 indicates we have statistically significant relationships. Few established relationships are statistically insignificant.

Develop the Logistic Classifier Using SciKit-Learn

In the next section, we use the Scikit-Learn package. We follow the same procedure as in other chapters. Listing 5-14 trains the model with default hyperparameters.

Listing 5-14. Develop the Logistic Classifier with Default Hyperparameters

```
from sklearn.linear_model import LogisticRegression
logreg = LogisticRegression()
logreg.fit(x_train,y_train)
```

Logistic Hyperparameter Optimization

After reviewing the contents of the preceding chapters, you might have noticed that hyperparameters play a significant role in model performance. Before finalizing machine learning models, ensure that a model has optimal hyperparameters. Listing 5-15 returns optimal hyperparameters.

Listing 5-15. Hyperparameter Optimization

```
from sklearn.model_selection import GridSearchCV
param_grid = {"dual":[False,True],
              "fit_intercept":[False,True],
              "max_iter":[1,10,100,1000],
              "penalty":("l1","l2"),
              "tol":[0.0001,0.001,0.01,1.0],
              "warm_start":[False,True]}
grid_model = GridSearchCV(estimator=logreg, param_grid=param_grid)
grid_model.fit(x_train,y_train)
print("Best score: ", grid_model.best_score_, "Best parameters: ",
grid_model.best_params_)
```

Best score: 0.7960609991158407 Best parameters: {'dual': False, 'fit_intercept': True, 'max_iter': 100, 'penalty': 'l2', 'tol': 0.0001, 'warm_start': False}

Listing 5-16 completes the model.

Listing 5-16. Finalize the Logistic Classifier

```
logreg = LogisticRegression(dual= False,
                            fit_intercept= True,
                            max_iter= 10,
                            n_jobs= -5,
                            penalty= 'l2',
                            tol=0.0001,
                            warm_start= False)
logreg.fit(x_train,y_train)
```

Predictions

Listing 5-17 obtains predicted values and, thereafter, creates a dataframe that shows actual values and predicted values (see Table 5-5).

Listing 5-17. Actual Values and Predicted Values

```
y_pred = logreg.predict(x_test)
y_pred  = pd.DataFrame({"Actual":y_test, "Predicted": y_predlogreg})
y_pred
```

Table 5-5. *Actual Values and Predicted Values*

	Actual	Predicted
661	1	1
122	0	0
113	0	0
14	1	1
529	0	0
...	...	...
476	1	0
482	0	0
230	1	1
527	0	0
380	0	0

Find the Intercept

Listing 5-18 finds the intercept.

Listing 5-18. Intercept

```
logreg.intercept_
array([-0.78763914])
```

Find the Estimated Coefficients

Listing 5-19 shows the parameters of our model.

Listing 5-19. Coefficients

```
logreg.coef_
array([[ 0.3097449 ,  1.06006236, -0.26057825,  0.06865213, -0.15816976,
         0.68419394,  0.29353764,  0.2396453 ]])
```

Evaluate the Logistic Classifier

After finalizing a binary classifier, we must look at how well the binary classifier classifies classes. This involves comparing the difference between actual and predicted labels. We begin with developing a confusion matrix.

Confusion Matrix

When the classifier makes predictions, it is prone to make errors. When dealing with a binary classification problem, there are two types of errors, namely, the Type I error (False Positive) and the Type II error (False Negative).

> *False positive (FP)*: Incorrectly predicting that an event took place. An example is predicting that a patient is diabetic when a patient is not diabetic.
>
> *False negative (FN)*: Incorrectly predicting an event that never took place. An example is predicting that a patient is not diabetic when a patient is diabetic.

Although the classifier is prone to making errors, it does get some predictions correct.

True positive (TP): Correctly predicting that an event took place. An example is correctly predicting that a patient is diabetic.

True negative (TN): Correctly predicting that an event never took place. An example is correctly predicting that a patient is not diabetic.

A confusion matrix takes four combinations of actual classes and predictions. A confusion matrix looks like Table 5-6, Listing 5-20, and Table 5-7.

Table 5-6. *Confusion Matrix*

	Predicted: No	Predicted: Yes
Actual: No	TN = 98	FP = 9
Actual: Yes	FN = 18	TP = 29

Listing 5-20. Confusion Matrix

```
cmatlogreg = pd.DataFrame(metrics.confusion_matrix(y_test,y_predlogreg),
index=["Actual: No","Actual: Yes"],
                          columns=("Predicted: No","Predicted: Yes"))
cmatlogreg
```

Table 5-7. *Confusion Matrix*

	Predicted: No	Predicted: Yes
Actual: No	98	9
Actual: Yes	18	29

Classification Report

A classification report is a detailed report that highlights how well the classifier performs. It encompasses key metrics such as accuracy, precision, F-1 support, recall, and prevalence. Calculations of these metrics come from the confusion matrix. Table 5-8 outlines key metrics in a classification report, and Listing 5-21 returns a classification report.

Table 5-8. Classification Report

Metric	Description
Precision	Determines how often the classifier is correct
Accuracy	Determines how often the classifier got predictions right
F1-support	Determines the mean value of precision and recall
Support	Determines the number of samples of the true response that lies in that class

Listing 5-21. Classification Report

```
creportlogreg = pd.DataFrame(metrics.classification_report(y_test,
y_predlogreg, output_dict=True)).transpose()
creportlogreg
```

Table 5-9 highlights that the logistic classifier is accurate 82% of the time. It also tells us that there is an imbalance in the data. The classifier is more precise when it predicts class 0 than when it predicts class 1.

Table 5-9. Classification Report

	precision	recall	f1-score	support
0	0.844828	0.915888	0.878924	107.000000
1	0.763158	0.617021	0.682353	47.000000
Accuracy	0.824675	0.824675	0.824675	0.824675
macro avg	0.803993	0.766455	0.780638	154.000000
weighted avg	0.819902	0.824675	0.818931	154.000000

ROC Curve

A receiver operating characteristics (ROC) curve is a tool used to find appropriate hyperparameter settings. It helps summarize the trade-off between the true positive rate and the false positive rate using different probability thresholds. On the x-axis lies the false positive rate, and on the y-axis lies the true positive rate. The closer the curve follows the left-side border than the top border of the ROC space, the more accurate it is. See Listing 5-22.

Listing 5-22. ROC Curve

```
y_predlogreg_proba = logreg.predict_proba(x_test)[::,1]
fprlogreg, tprlogreg, _ = metrics.roc_curve(y_test,y_predlogreg_proba)
auclogreg = metrics.roc_auc_score(y_test, y_predlogreg_proba)
plt.plot(fprlogreg, tprlogreg, label="AUC: "+str(auclogreg), color="navy")
plt.plot([0,1],[0,1],color="red")
plt.xlim([0.00,1.01])
plt.ylim([0.00,1.01])
plt.xlabel("Specificity")
plt.ylabel("Sensitivity")
plt.legend(loc=4)
plt.show()
```

The curve in Figure 5-7 follows the left-side border for a while; then it smoothly bends to the top border of the ROC space. The curve slowly approaches the 45-degree line. This means that the logistic classifier is not accurate as we would want it to be.

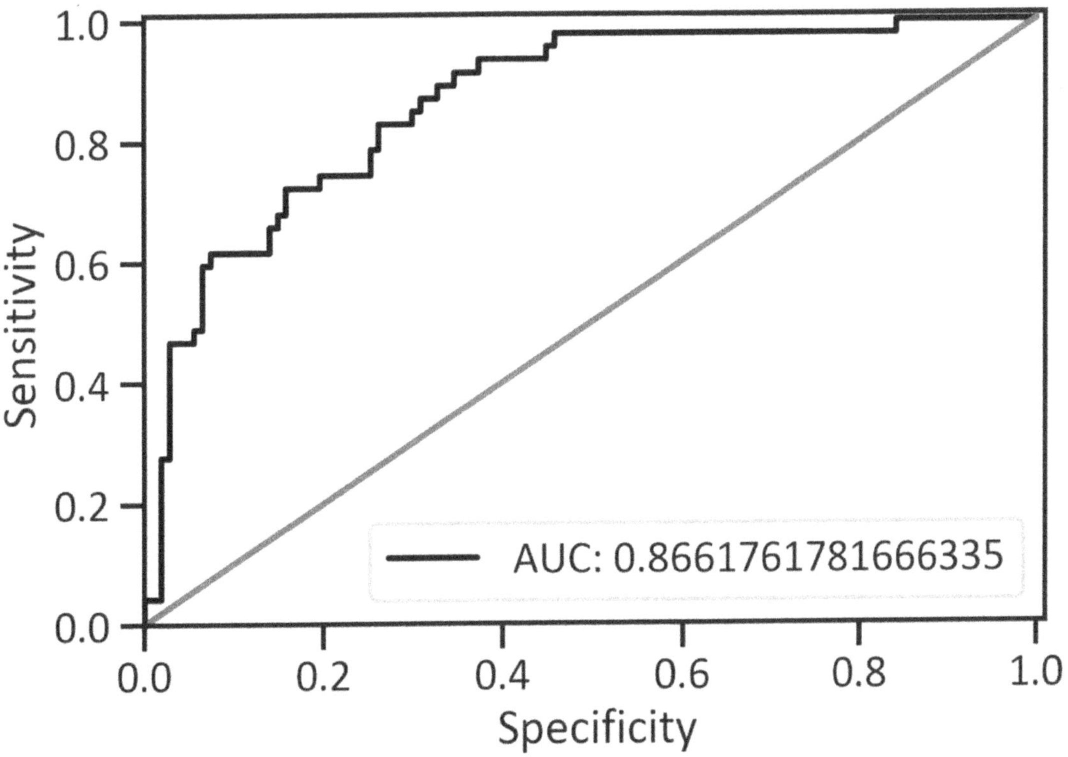

Figure 5-7. *ROC curve*

Area Under the Curve

The area under the curve (AUC) is also recognized as the index of accuracy. It consists of performance metrics for an ROC curve. It indicates how the classifier distinguishes between classes. The closer the area under the curve score is to 1, the better the predictive power the model has. The AUC score is 0.87. Using the 80/20 rule, the classifier is skillful in distinguishing classes between actual classes and predicted classes.

Precision Recall Curve

We use the precision-recall curve to show the trade-offs between precision and recall across different thresholds. Ideally, we want a curve that straightforwardly moves to the top right and sharply bends horizontally. Such a curve has high precision and high recall. This means that the binary classifier gets all the predictions correct. See Listing 5-23.

Listing 5-23. Precision-Recall Curve

```
precisionlogreg, recalllogreg, thresholdlogreg = metrics.precision_recall_
curve(y_test,y_predlogreg)
apslogreg = metrics.roc_auc_score(y_test,y_predlogreg)
plt.plot(precisionlogreg, recalllogreg, label="aps: "+str(apslogreg),
color="navy",alpha=0.8)
plt.axhline(y=0.5,color="red",alpha=0.8)
plt.xlabel("Precision")
plt.ylabel("Recall")
plt.legend(loc=4)
plt.show()
```

The curve in Figure 5-8 does not approach the top-right border; rather, it slowly declines to the bottom right. This indicates that there is an overlap between patients with diabetes and those without diabetes. The precision and recall are not high enough.

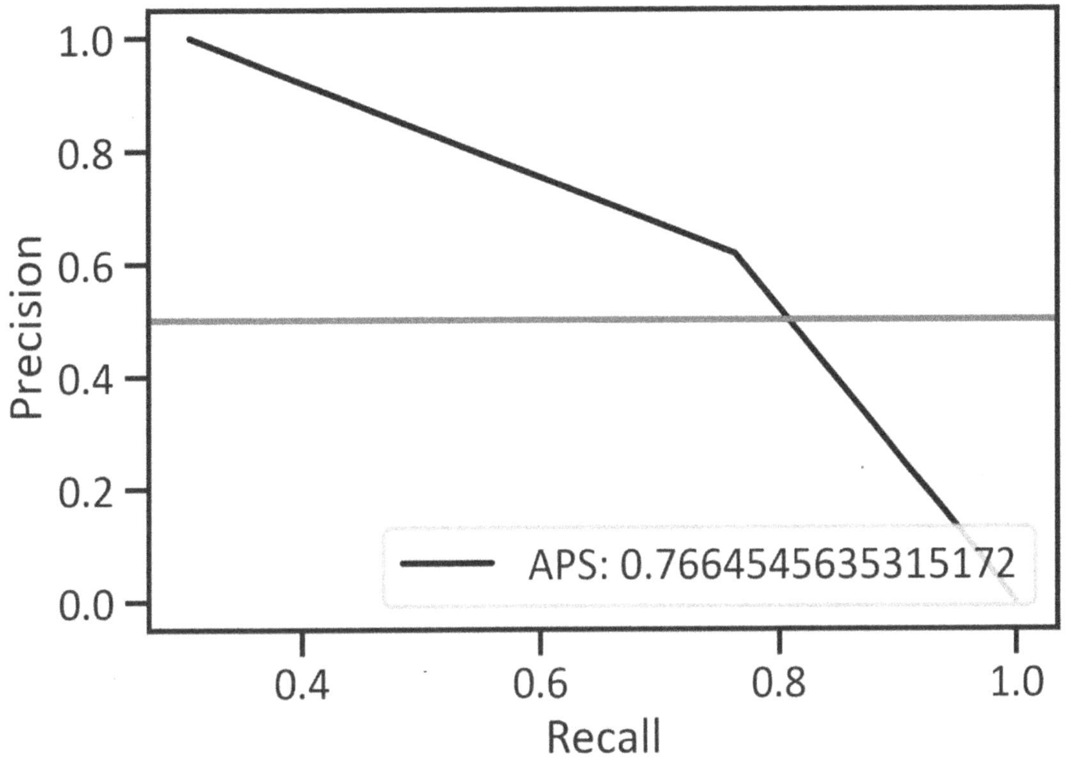

Figure 5-8. *Precision-recall curve*

Find the Average Precision Score

To make sense of a precision-recall curve, we unbundle the central tendency. An average precision score (APS) is the arithmetic average of the precision scores inside the precision-recall curve. The classifier is precise 76% of the time.

Learning Curve

A learning curve finds whether a model has a low bias error term (under-fitting) and high variance (over-fitting). It depicts a model's learning process over time. This curve enables us to determine the extent to which we can benefit from increased data points in the training data and to find out whether the estimator suffers variance error or bias. There are two types of learning curves, namely, training learning curve (calculated from the training data) and validation learning curve (calculated using the validation data). A learning curve has two axes: the training set size on the x-axis and the accuracy score on the y-axis. Figure 5-9 shows the classifier's learning process. See also Listing 5-24.

Listing 5-24. Learning Curve

```
trainsizelogreg, trainscorelogreg, testscorelogreg = learning_curve(logreg,
x, y, cv=5, n_jobs=5, train_sizes=np.linspace(0.1,1.0,50))
trainscorelogreg_mean = np.mean(trainscorelogreg,axis=1)
testscorelogreg_mean = np.mean(testscorelogreg,axis=1)
plt.plot(trainsizelogreg,trainscorelogreg_mean,color="red", label="Training
score", alpha=0.8)
plt.plot(trainsizelogreg,testscorelogreg_mean,color="navy", label="Cross
validation score", alpha=0.8)
plt.xlabel("Training set size")
plt.ylabel("Accuracy")
plt.legend(loc=4)
plt.show()
```

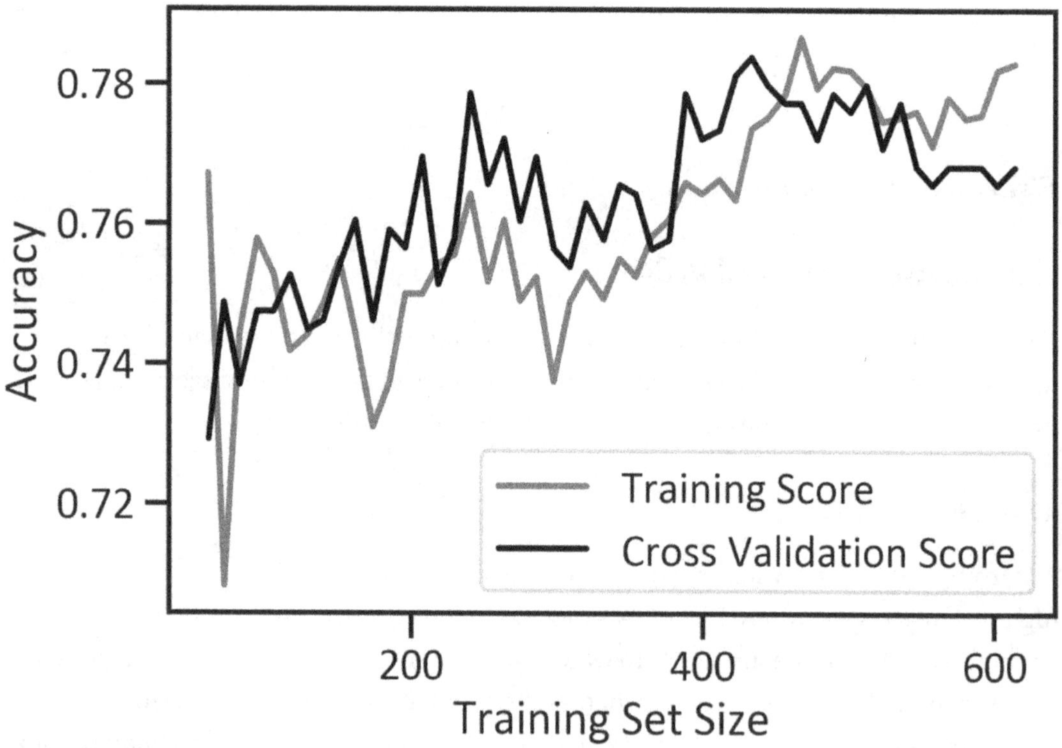

Figure 5-9. *Learning curve*

At the beginning phase of the learning process, the logistic classifier over-fits. However, as the training set size increases, the classifier over-fits less. As the classifier approaches the 400th data point, the cross-validation accuracy score starts to decrease. Adding more data points will not increase generalization. You will learn more about how to interpret a learning curve in the subsequent chapters.

Conclusion

This chapter introduced the nonparametric method. It assessed two logistic classifiers developed using Statsmodels and SciKit-Learn. The second mode shows the characteristics of a well-behaved model. The model is correct 85% of the time. Ideally, we want a model that is correct 100% of the time. There are several ways of improving the performance of the logistic classifier, such as dimension reduction and regularization. In the SciKit-Learn package, there is the logistic classifier with built-in cross validation and a ridge classifier. These models can help improve the performance of the classifier.

CHAPTER 6

Dimension Reduction and Multivariate Analysis Using Linear Discriminant Analysis

The preceding chapter presented a classification method known as logistic regression. It solves binary classification problems. Multinomial logistic regression (MLR) is an extension of logistic regression using the Softmax function; instead of the Sigmoid function, it applies the cross-entropy loss function. It is a form of logistic regression used to predict a target variable with more than two classes. It differs from linear discriminant analysis (LDA) in the sense that MLR does not assume the data comes from a normal distribution. LDA comes from a linear family; it assumes normality and linearity.

We use LDA to solve binary class problems and multiclass classification problems. This chapter introduces linear classification and discriminant analysis. It reveals a way of using LDA to condense data to fewer dimensions and estimate categorical variables using continuous variables.

This model works like models concealed in the preceding chapters, like linear regression and logistic regression. It is also similar to analysis of covariance (ANOVA). ANOVA estimates continuous variables using categorical variables. LDA reasonably assumes that the covariance structure of the groups is alike. If the covariance structure groups are unequal, then use the quadratic discriminant analysis. An extension of the LDA model is known as Fisher's discriminant analysis (FDA). FDA does not perform classification, but features obtained after transformation can be used for classification.

© Tshepo Chris Nokeri 2021
T. C. Nokeri, *Data Science Revealed*, https://doi.org/10.1007/978-1-4842-6870-4_6

Both logistic regression and LDA estimate a categorical dependent variable. What differentiates LDA from logistic regression has more assumptions. It is also like principal component analysis and factor analysis in the sense it looks for linear combinations that best explain the data. LDA works best when there is a large sample.

It applies the discriminant function to allocate a group of continuous variables into categories by finding the linear combinations of variables and maximizing the difference between them. Likewise, it applies Fisher's linear discriminant rule to maximize the ratio between the within-class scatter matrix and between-class scatter matrix and to discover the linear combinations in groups.

Last, it estimates categorical variables using only continuous variables. If independent variables are categorical and the dependent variable is categorical, use an alternative model recognized as discriminant correspondence analysis. Given its inheritance from the linear family, test the model against assumptions of linearity and normality. LDA serves two purposes: dimension reduction and classification.

Dimension Reduction

We use LDA, which is similar to PCA, for dimension reduction. It discovers variance explained in the data and reduces data into fewer dimensions. Remember to use eigenvalues to find the source of considerable variation in the data. An eigenvalue with the uppermost loading is the first function, the one with the second-highest loading is the second function, and so forth. We eliminate the factors with excessive eigenvalues (extreme variability) based on some criteria.

Classification

LDA solves both binary and multiclass classification problems. It applies a linear classifier to assign variables to a class. Given that the model applies a linear classifier, it inherits strict assumptions of linearity and normality.

Assumptions

LDA makes strong assumptions about the structure of the data. We consider the LDA classifier reliable if it satisfies the assumptions highlighted in Table 6-1.

Table 6-1. *LDA Assumptions*

Assumption	Description
Multivariate normality	There must be multivariate normality in each grouping of variables; independent variables must follow a normal distribution.
Covariance	Variances among group variables are the same across levels of predictors. When covariance is equal, one can make use of linear discriminant analysis. Otherwise, use quadratic discriminant analysis.
Correlation	Variables must not be highly correlated.
Random sampling	Participants must be selected using random sampling. A participant's score must be independent of variables.

Assuming there are C classes, let the mean vector of the class be the number of samples within classes.

Equation 6-1 is for a within-class scatter matrix.

$$S_w = \sum_{i=1}^{C}\sum_{j=1}^{M_i}\left(y_j - \mu_i\right)\left(y_j - \mu_i\right)^T \qquad \text{(Equation 6-1)}$$

Equation 6-2 is for a between-class scatter matrix. Equation 6-3 is within scatter matrix.

$$S_b = \sum_{i=1}^{C}\left(y_j - \mu_i\right)\left(y_j - \mu_i\right)^T \qquad \text{(Equation 6-2)}$$

$$\mu = \frac{1}{C}\sum_{i=1}^{C}\mu_i\left(mean\ of\ the\ entire\ dataset\right) \qquad \text{(Equation 6-3)}$$

It performs a transformation that maximizes the between-class scatter matrix while minimizing the within-scatter. See Equation 6-4.

$$maximize = \frac{\det\left(S_w\right)}{\det\left(S_b\right)} \qquad \text{(Equation 6-4)}$$

This transformation keeps class separability while reducing the variation due to sources other than identity. The linear transformation is given by a matrix *U* whose columns are the eigenvectors of $S_w^{-1}S_b$.

Basically, it computes the d-dimensional mean vectors, followed by the scatter matrices and eigenvectors and equivalent eigenvalues for the scatter matrices. Thereafter, the model class the eigenvalues and select larger eigenvalues to form a $d{\times}k$ dimensional matrix and convert the samples onto the new subspace. The end result is a class separation.

Develop the LDA Classifier

The example data was retrieved from ML Data.[1] We want to predict the quality of a car based on a set of independent variables. See Listing 6-1.

Listing 6-1. Develop the Model with Default Hyperparameters

```
from sklearn.discriminant_analysis import LinearDiscriminantAnalysis
LDA = LinearDiscriminantAnalysis()
LDA.fit(x_train,y_train)
```

LDA Hyperparameter Optimization

Tuning LDA hyperparameters is not as tedious as tuning hyperparameters of other classification models. Table 6-2 highlights important hyperparameters.

Table 6-2. *Tunable Hyperparameters*

Parameters	Description
`n_component`	Determines the number of components.
`shrinkage`	Determines the automatic shrinkage using the Ledoit-Wolf lemma or specifies the value.
`solver`	Determines the solver to use; use `svd` for single-value decomposition, `lsqr` for least squares solutions, and `eigen` for Eigenvalue decomposition by default
`store_covariance`	Determines whether to store covariance
`tol`	Determines the tolerance for the optimization

[1] https://www.mldata.io/dataset-details/cars/

Listing 6-2 finds the hyperparameters that yield optimal model performance.

Listing 6-2. Hyperparameter Optimizations

```
param_gridLDA = {"n_components":[1,2,3,4,5,6,7,8,9],
                 "solver":("svd", "lsqr", "eigen"),
                 "store_covariance":[False,True],
                 "tol":[0.0001,0.001,0.01,1.0]}
grid_modelLDA = GridSearchCV(estimator=LDA, param_grid= param_gridLDA)
grid_modelLDA.fit(x_train,y_train)
print("Best score: ", grid_modelLDA.best_score_, "Best parameters: ",
grid_modelLDA.best_params_)
```

Best score: 0.7622550979608157 Best parameters: {'n_components': 1, 'solver': 'svd', 'store_covariance': False, 'tol': 0.0001}

Listing 6-3 completes the model using default hyperparameters.

Listing 6-3. Finalize the LDA Classifier

```
LDA = LinearDiscriminantAnalysis(n_components= 1,
                                 solver= 'svd',
                                 store_covariance= False,
                                 tol= 0.0001)
LDA.fit(x_train,y_train)
```

Listing 6-4 transforms the variables.

Listing 6-4. Create New Features

```
features_new = LDA.transform(x_train)
```

Listing 6-5 estimates the percentage of variance explained by each of the selected components.

Listing 6-5. Explained Variance Ratio

```
print(LDA.explained_variance_ratio_)
[1.]
```

Each of the selected components explains 100% of the variance. Listing 6-6 prints the original number of features and the number of reduced features.

Listing 6-6. The Number of Original and Reduced Features

```
print('Original feature #:', x_train.shape[1])
print('Reduced feature #:', features_new.shape[1])
Original feature #: 7
Reduced feature #: 1
```

Initially, we had eight features; there are now seven features.

Predictions

After finalizing the classifier, the subsequent step involves comparing the actual values of the dependent variable and predicted values of the variable. Listing 6-7 tabulates actual values and predicted values (see Table 6-3).

Listing 6-7. Actual Values and Predicted Values

```
y_predLDA = LDA.predict(x_test)
pd.DataFrame({"Actual": y_test,"Predicted": y_predLDA})
```

Table 6-3. *Actual Values and Predicted Values*

	Actual	Predicted
661	1	1
122	0	0
113	0	0
14	1	1
529	0	0
...	...	...
476	1	0
482	0	0
230	1	1
527	0	0
380	0	0

Table 6-3 does not tell us much about how the classifier performs.

Evaluate the LDA Classifier

To understand how the LDA classifier performs, we must compare actual classes and predicted classes side by side.

Confusion Matrix

We commonly recognize a confusion matrix as the error matrix. It is a 2×2 matrix that counts actual labels and predicted labels. Listing 6-8 tabulates the confusion matrix (see Table 6-4).

Listing 6-8. Confusion Matrix

```
cmatLDA = pd.DataFrame(metrics.confusion_matrix(y_test,y_predLDA),
index=["Actual: No","Actual: Yes"],
                              columns=("Predicted: No","Predicted: Yes"))
cmatLDA
```

Table 6-4. *Confusion Matrix*

	Predicted: No	Predicted: Yes
Actual: No	263	0
Actual: Yes	0	263

Table 6-4 highlights values similar to the logistic classifier. We expect to find similar performance results.

Classification Report

Table 6-5 highlights important classification evaluation metrics. It gives us an overview of how well the classifier performs. See Listing 6-9.

Listing 6-9. Classification Report

```
creportLDA = pd.DataFrame(metrics.classification_report(y_test,y_predLDA,
output_dict=True)).transpose()
creportLDA
```

Table 6-5. *Classification Report*

	precision	recall	f1-score	support
0	1.0	1.0	1.0	263.0
1	1.0	1.0	1.0	83.0
accuracy	1.0	1.0	1.0	1.0
macro avg	1.0	1.0	1.0	346.0
weighted avg	1.0	1.0	1.0	346.0

The LDA classifier shows the optimal model performance. It is accurate and precise 100% of the time.

ROC Curve

Listing 6-10 applies the `roc_curve()` and `roc_auc_score()` method to depict actual classes and probabilities of each class to develop the curve (see Figure 6-1).

Listing 6-10. ROC Curve

```
y_predLDA_probaLDA = LDA.predict_proba(x_test)[::,1]
fprLDA, tprLDA, _ = metrics.roc_curve(y_test,y_predLDA_probaLDA)
aucLDA = metrics.roc_auc_score(y_test,y_predLDA_probaLDA)
plt.plot(fprLDA, tprLDA,label="AUC: " + str(aucLDA),color="navy")
plt.plot([0,1], [0,1],color="red")
plt.xlim([0.00,1.01])
```

```
plt.ylim([0.00,1.01])
plt.xlabel("Sensitivity")
plt.ylabel("Specificity")
plt.legend(loc=4)
plt.show()
```

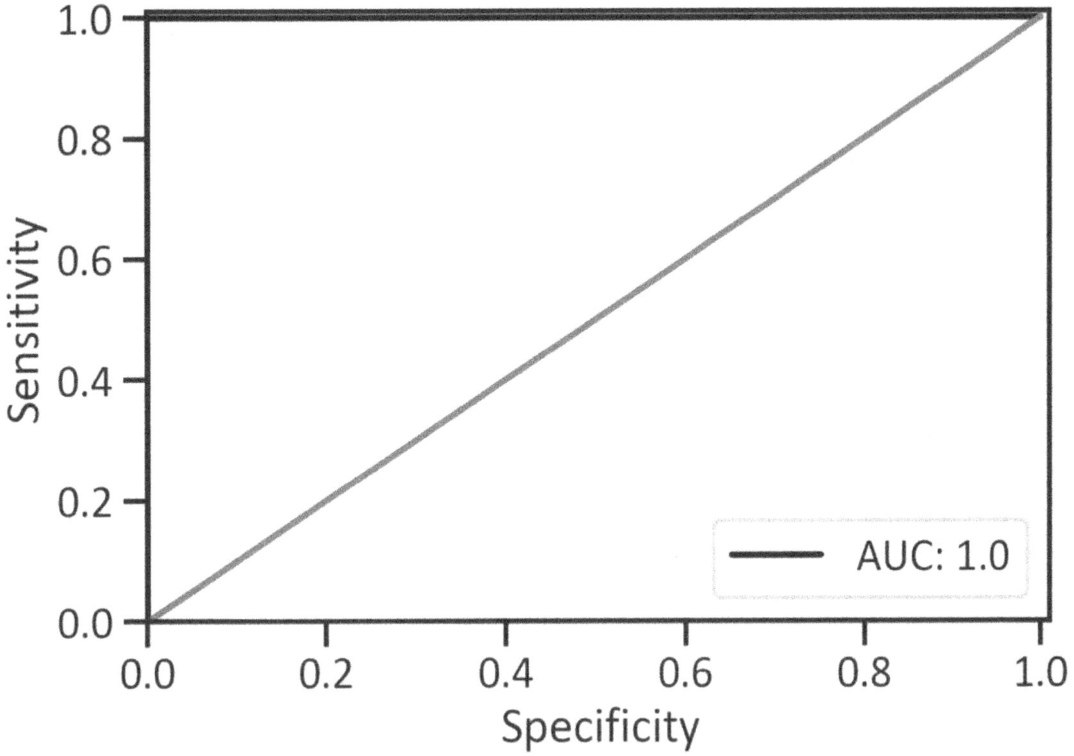

Figure 6-1. *ROC curve*

Figure 6-1 shows that the AUC score is 1.0. The LDA classifier shows how the optimal model performs.

Precision-Recall Curve

Listing 6-11 produces a curve that tells us about the relevance and precision of the LDA classifier when it predicts classes. We use a precision-recall curve.

Listing 6-11. Precision-Recall Curve

```
precisionLDA, recallLDA, thresholdLDA = metrics.precision_recall_curve(
y_test,y_predLDA)
apsLDA = metrics.roc_auc_score(y_test,y_predLDA)
plt.plot(precisionLDA, recallLDA, label="APS: "+str(apsLDA),color="navy")
plt.axhline(y=0.5,color="red",alpha=0.8)
plt.xlabel("Recall")
plt.ylabel("Precision")
plt.legend(loc=4)
plt.show()
```

Figure 6-2 shows characteristics of a well-behaved precision-recall curve.

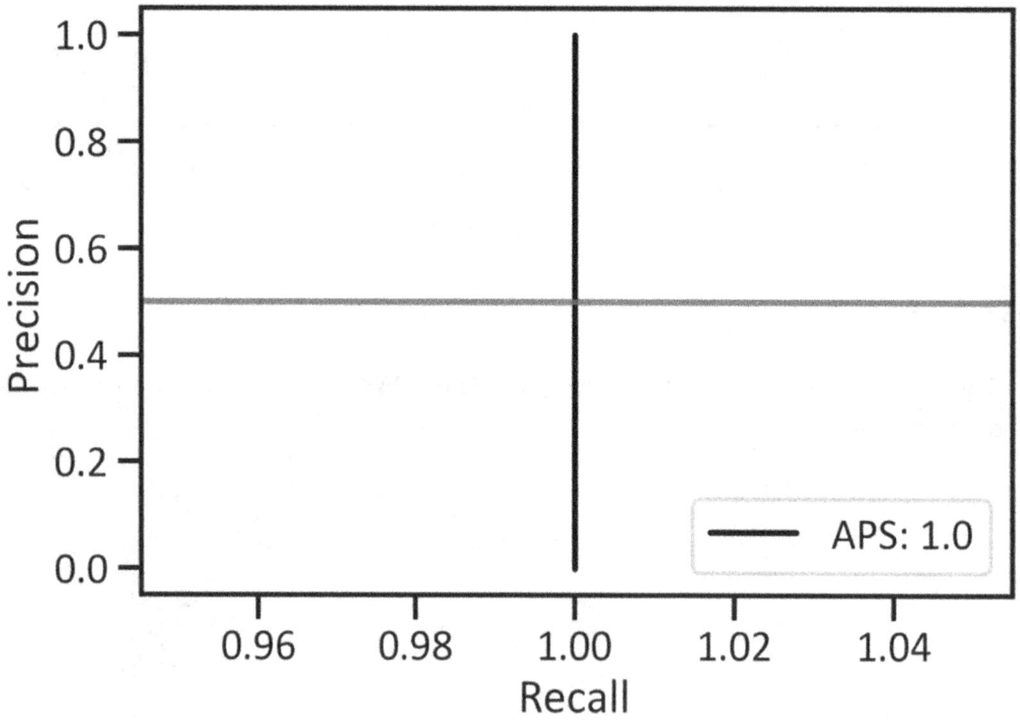

Figure 6-2. *Precision-recall curve*

Learning Curve

Listing 6-12 and Figure 6-3 depict the LDA classifier's learning process.

Listing 6-12. Learning Curve

```
trainsizeLDA, trainscoreLDA, testscoreLDA = learning_curve(LDA, x, y, cv=5,
n_jobs=5, train_sizes=np.linspace(0.1,1.0,50))
trainscoreLDA_mean = np.mean(trainscoreLDA,axis=1)
testscoreLDA_mean = np.mean(testscoreLDA,axis=1)
plt.plot(trainsizeLDA,trainscoreLDA_mean,color="red", label="Training Score")
plt.plot(trainsizeLDA,testscoreLDA_mean,color="navy", label="Cross
Validation Score")
plt.xlabel("Training Set Size")
plt.ylabel("Accuracy")
plt.legend(loc=4)
plt.show()
```

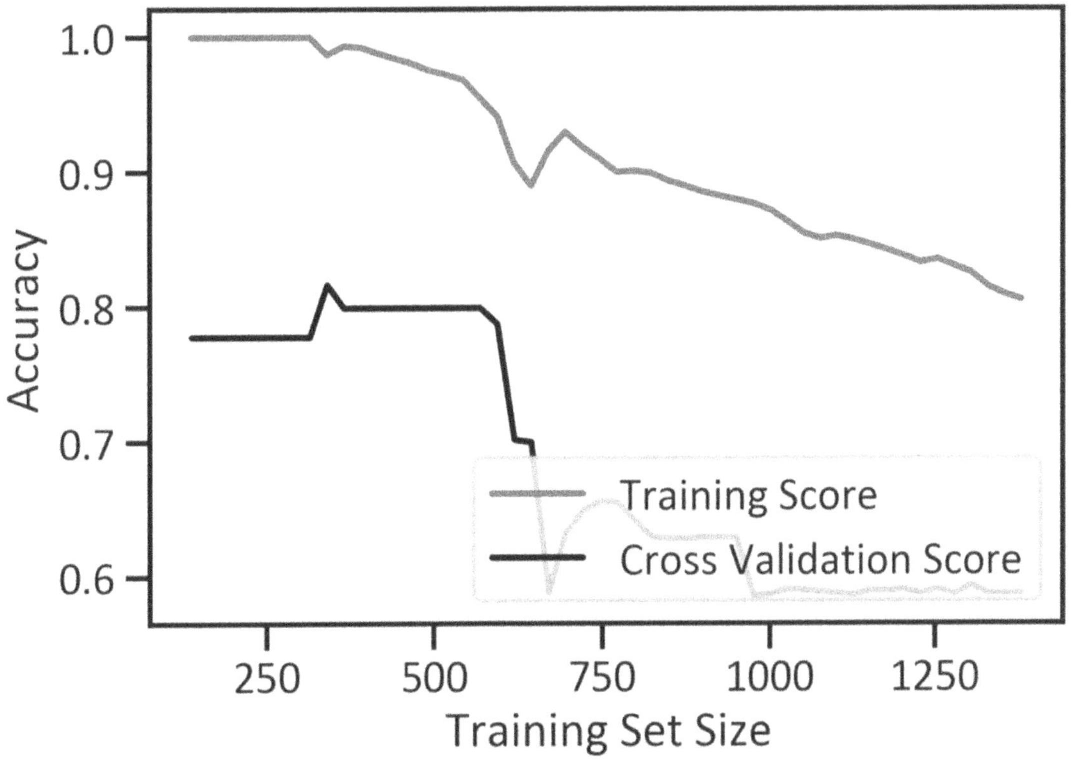

Figure 6-3. *Learning curve*

Figure 6-3 conveys that the classifier's training and cross-validation accuracy is stable for the first 550 data points, but as we increase the training set, the accuracy drops.

Conclusion

This chapter explored discriminant analysis. It focused primarily on linear discriminant analysis and illustrated ways of using it for dimension reduction linear classification. The model used a single decomposition solver, which was restricted from storing the covariance and tolerance of the optimization set to 0.0001. It shows the optimal model performance.

CHAPTER 7

Finding Hyperplanes Using Support Vectors

The preceding chapter presented a linear classification model called linear discriminant analysis (LDA), which distributes groups equally when covariance matrices are equivalent. Although the classifier is one of the optimum linear classification models, it has its limits. Foremost, we cannot estimate the dependent variable using a categorical variable. Second, we train and test the model under strict assumptions of normality. This chapter brings together an alternative linear classification model called support vector machine (SVM). It is part of the ensemble family; it estimates either a continuous variable or a categorical variable. It applies a kernel function to transform data in such a way that a hyperplane best fits the data. Unlike LDA, SVM makes no assumptions about the underlying structure of the data.

Support Vector Machine

SVM typically assumes that the data points are linearly indistinguishable in lower dimensions and linearly distinguishable in higher dimensions. It finds an optimal hyperplane to take full advantage of the margin and simplify nonlinear problems. First, it efficiently captures data and develops a new dimensional vector space. Thereafter, it develops a linear boundary of the new vector space into two classes. It positions a data point either above or underneath a hyperplane, resulting in classification. It uses a kernel function to estimate a hyperplane with an extreme margin. The primary kernel functions are the linear function, polynomial function, and sigmoid function. Except if you are analyzing high-dimensional data, there is absolutely no need to study these functions extensively. You can always use hyperparameter optimization to find an optimal function.

© Tshepo Chris Nokeri 2021
T. C. Nokeri, *Data Science Revealed*, https://doi.org/10.1007/978-1-4842-6870-4_7

Support Vectors

Support vectors are data points next to the hyperplane. They determine the position of the hyperplane. Furthermore, when we eliminate these data points, the position of a hyperplane changes.

Hyperplane

A hyperplane is a line that splits into two categories (each class falls on either side). It serves as a boundary between the two sides. Positioning an optimal hyperplane is challenging when using real-world data. SVM transforms the data into higher dimensions through a procedure called *kerneling*. After transforming the data, it does not develop a straight line but a plane. It applies a kernel function to transform the data, so it positions a hyperplane at the proper location. Figure 7-1 portrays a hyperplane.

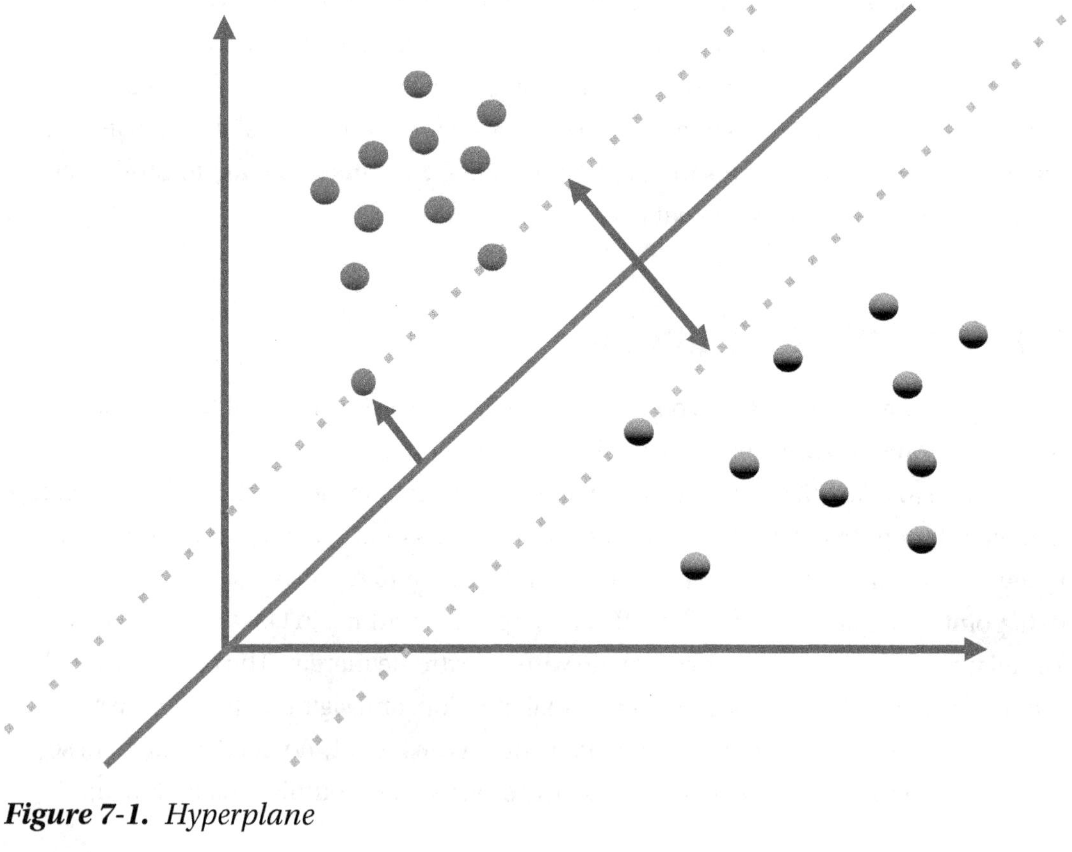

Figure 7-1. *Hyperplane*

Figure 7-1 shows the formula of the straight line expressed in Equation 7-1.

$$WT\ x + b = 0 \qquad \text{(Equation 7-1)}$$

Here, *W* represents the slope of the line, *x* represents the input vector, and *b* represents bias. The two lines (highlighted in orange) pass through the support vectors and support the best plane. A decent hyperplane has an extreme margin for the support vectors.

It figures out how to position a hyperplane during training by using an optimization procedure that enlarges the separation. We do not add hyperplanes manually. It applies a technique recognized as a kernel trick to transform the data. The technique enables it to enlarge the dimensional vector space with less computing power.

Support Vector Classifier

A support vector classifier (SVC), also known as a *margin classifier*, permits certain data points to be on the inappropriate side of the hyperplane. If the data is not directly distinguishable, it includes slack variables to permit the violation of constraints by allowing certain data points to fall inside the margin; in addition, it punishes them. It limits slack variables to non-negative values. If slack variables are over 0, it sets aside constraints by positioning a plane next to the data points instead of the margin by selecting a lambda. Figure 7-2 shows how the SVC works.

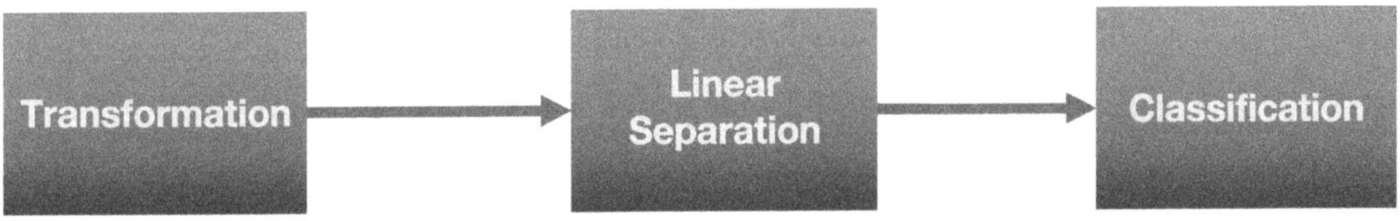

***Figure 7-2.** LinearSVC workflow*

There are three basic tasks the LinearSVC classifier performs.

- *Transformation*: Changing an input space into higher-dimensional vectors space
- *Linear separation or straight division*: Realizing a hyperplane with the most margin
- *Classification*: Allocating data points to a class

Develop the LinearSVC Classifier

We obtained the example data from Kaggle.[1] Listing 7-1 completes the classifier.

Listing 7-1. Finalize the LinearSVC Classifier

```
from sklearn.svm import LinearSVC
lsvc = LinearSVC()
lsvc.fit(x_train,y_train)
```

LinearSVC Hyperparameter Optimization

Table 7-1 highlights key hyperparameters.

Table 7-1. *Tunable Hyperparameters*

Parameter	Description
`Dual`	Determines whether the model must solve dual problems
`fit_intercept`	Determines whether the model must calculate the intercept.
`max_iter`	Determines the maximum number of iterations.
`Penalty`	Determines the regularization method to use. Available penalty terms include l1 penalty (lasso) and l2 penalty (ridge).
`Tol`	Determines the tolerance for the optimization.
`Loss`	Determines the type of function. Available loss functions include hinge and squared hinge.

Listing 7-2 creates a dictionary with group parameters and a specific range of values we expect to find the best cross-validation score from and then applies the `GridSearchCV()` method to develop a grid model and find hyperparameters with the best score.

[1]Visit `https://www.kaggle.com/uciml/pima-indians-diabetes-database` to download the data.

Listing 7-2. Hyperparameter Optimization

```
param_gridlsvc = {"dual":[False,True],
                  "fit_intercept":[False,True],
                  "max_iter":[1,10,100,100],
                  "penalty":("l1","l2"),
                  "tol":[0.0001,0.001,0.01,1.0],
                  "loss":("hinge", "squared_hinge")}
grid_modellsvc = GridSearchCV(estimator=lsvc, param_grid=param_gridlsvc)
grid_modellsvc.fit(x_train,y_train)
print("Best score: ", grid_modellsvc.best_score_, "Best parameters: ",
grid_modellsvc.best_params_)
```

Best score: 0.7671331467413035 Best parameters: {'dual': False, 'fit_intercept': True, 'loss': 'squared_hinge', 'max_iter': 100, 'penalty': 'l1', 'tol': 0.001}

Finalize the LinearSVC Classifier

The previous findings suggest that we use the squared hinge function and control for bias and variance using the L1 penalty term. Listing 7-3 completes the LinearSVC classifier.

Listing 7-3. Finalize the LinearSVC Classifier

```
lsvc = LinearSVC(dual= False,
                 fit_intercept= True,
                 max_iter= 100,
                 penalty= 'l1',
                 tol= 0.001,
                 loss="squared_hinge")
lsvc.fit(x_train,y_train)
```

Evaluate the LinearSVC Classifier

Listing 7-4 tabulates the actual classes and predicted classes side by side. See also Table 7-2.

Listing 7-4. Actual Values and Predicted Values

```
y_predlsvc = lsvc.predict(x_test)
pd.DataFrame({"Actual":y_test,"Predicted":y_predlsvc})
```

Table 7-2. *Actual Values and Predicted Values*

	Actual	Predicted
661	1	1
122	0	0
113	0	0
14	1	1
529	0	0
...	...	...
476	1	0
482	0	0
230	1	1
527	0	0
380	0	0

Confusion Matrix

Listing 7-5 and Table 7-3 give us an abstract report about the classifier's performance.

Listing 7-5. Confusion Matrix

```
cmatlsvc = pd.DataFrame(metrics.confusion_matrix(y_test,y_predlsvc),
                        index=["Actual: No","Actual: Yes"],
                        columns=("Predicted: No","Predicted: Yes"))
cmatlsvc
```

Table 7-3. *Confusion Matrix*

	Predicted: No	Predicted: Yes
Actual: No	98	9
Actual: Yes	18	29

Classification Report

Listing 7-6 and Table 7-4 show how reliable the LinearSVC classifier is. It provides the model's accuracy score, precision score, recall, and other key evaluation metrics.

Listing 7-6. Classification Report

```
creportlsvc = pd.DataFrame(metrics.classification_report(y_test,y_predlsvc,
output_dict=True)).transpose()
creportlsvc
```

Table 7-4. *Classification Report*

	precision	recall	f1-score	support
0	0.844828	0.915888	0.878924	107.000000
1	0.763158	0.617021	0.682353	47.000000
accuracy	0.824675	0.824675	0.824675	0.824675
macro avg	0.803993	0.766455	0.780638	154.000000
weighted avg	0.819902	0.824675	0.818931	154.000000

The LinearSVC classifier is accurate 82% of the time. It is precise 84% of the time when predicting class 0 and 76% of the time when predicting class 1.

Learning Curve

Listing 7-7 produces a curve that depicts the progression of the LinearSVC classifier's accuracy as it learns the training data (see Figure 7-3).

Listing 7-7. Learning Curve

```
trainsizelsvc, trainscorelsvc, testscorelsvc = learning_curve(lsvc, x, y,
cv=5, n_jobs=5, train_sizes=np.linspace(0.1,1.0,50))
trainscorelsvc_mean = np.mean(trainscorelsvc,axis=1)
testscorelsvc_mean = np.mean(testscorelsvc,axis=1)
plt.plot(trainsizelsvc,trainscorelsvc_mean,color="red", label="Training Score")
plt.plot(trainsizelsvc,testscorelsvc_mean,color="navy", label="Cross
Validation Score")
plt.xlabel("Training Set Size")
plt.ylabel("Accuracy")
plt.legend(loc=4)
plt.show()
```

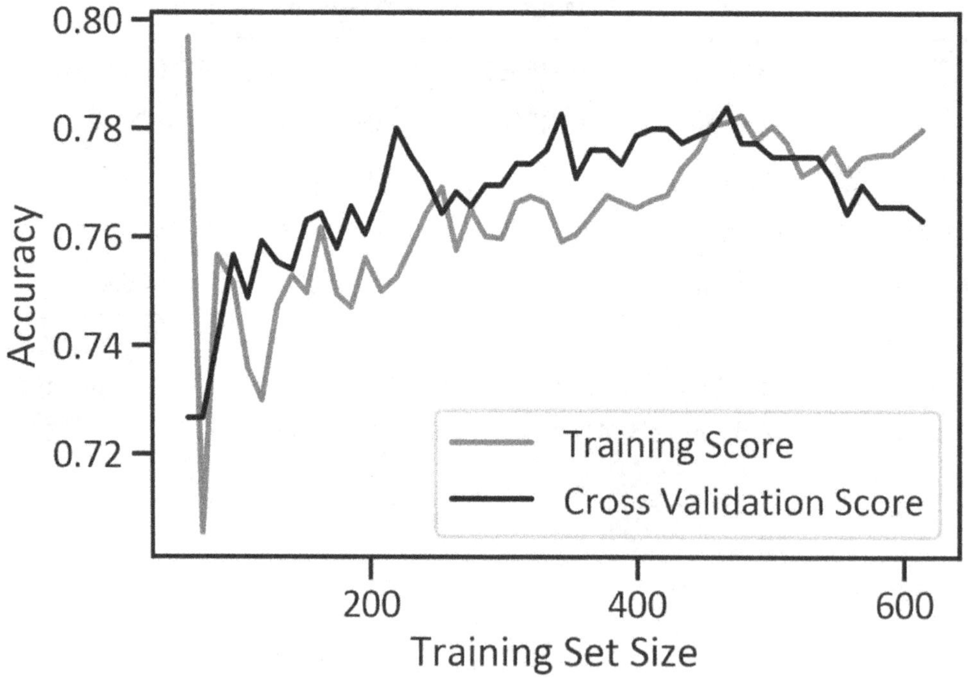

Figure 7-3. *Learning curve*

In the first phase of the learning process, the training accuracy score sharply declined to an all-time low of about 70%, followed by an upward rally. As the classifier reached the 200th data point, the training accuracy score improved. At most, the training accuracy score is below the cross-validation accuracy score. Last, as the classifier is about to reach the 600th data point, the training accuracy score surpasses the cross-validation accuracy score.

Conclusion

This chapter presented how SVM changes the input space into a higher-dimensional vector space and creates a hyperplane and classifies variables. We configured the LinearSVC classifier with a squared hinge loss function and an L1 penalty term. Remember, we do not develop ROC curve and precision-recall curve as the classifier lacks probabilistic importance. We rely upon the classification report and learning curve to comprehend the underlying performance of the classifier.

CHAPTER 8

Classification Using Decision Trees

This chapter presents the most widespread ensemble method, the decision tree. A decision tree classifier estimates a categorical dependent variable or a continuous dependent. It solves binary and multiclass classification problems. We base the model on a tree-like structure. It breaks down the data into small, manageable chunks while incrementally developing a decision tree. The outcome is a tree-like structure with decision nodes and leaf nodes. We consider it a greedy model since its primary concern is to reduce the training time while maximizing information gain.

To get the best out of decision trees, knowing the data and partitions are required in advance. We construct a decision tree by recursive partitioning the training data into decision nodes. Thereafter, we examine variables and branch cases grounded on the findings of the examination.

It selects variables with excessive predictive power, less impurity, and low entropy. It also explores the paramount variable to decrease impurity. The most important thing is the purity of the leaves after partitioning; it discovers a tree with the lowest entropy on the nodes. In summary, the decision tree classifier serves two purposes: separating data into different subsets and shortening the branches of a decision tree. It limits the depth of the tree.

Entropy

Entropy represents an estimate of randomness or uncertainty. The lower the entropy, the less uniform the distribution is, and the more the nodes. It shows the level of disorder in the data. It estimates the homogeneity of samples in a node. If the variables are

© Tshepo Chris Nokeri 2021
T. C. Nokeri, *Data Science Revealed*, https://doi.org/10.1007/978-1-4842-6870-4_8

homogenous, then the entropy is zero. If we equally divide samples, then entropy is 1. We estimate entropy using Equation 8-1.

$$Entropy = p(A)\ log\big(p(A)\big) - p(B) log\big(p(B)\big) \qquad \text{(Equation 8-1)}$$

Here, *p* is the prop or ratio of a category. We use entropy to group comparative data groups into comparable classes. High entropy shows that we have highly disordered data. A low entropy shows the data is well-organized. The model assumes that a less impure node requires less information to describe it, and a more impure node requires more information.

A decision tree estimates the entropy of the node, and it then estimates the entropy of each node of the partition. Thereafter, it weighs averages of the subnodes and estimate information gain and selects the node with the highest information gain for partitioning. Organizing data for decision tree classification involves partitioning the data and group samples together in classes they belong to and maximizing the purity of the groups each time it develops a node.

Information Gain

Information gain refers to information that increases the level of certainty after a partition. It is the opposite of entropy. If weighted entropy decreases, then information gain increases, and vice versa. In summary, the decision tree classifier finds variables that return the most noteworthy information gain. To calculate information gain, we use Equation 8-2.

$$\begin{aligned} Information\ gain = &(Entropy\ before\ the\ split) - \\ &(Weighted\ entropy\ after\ the\ split) \end{aligned} \qquad \text{(Equation 8-2)}$$

Structure of a Decision Tree

As mentioned, a decision tree is a flowchart-based tree-like structure. A decision tree estimates a sin curve using a bunch of `if-then-else` decision rules. The complexity of the decision rules depends on the depth of the tree. A decision tree encompasses decision nodes and leaf nodes. The uppermost decision node corresponds to the root node (the noteworthy independent variable). A leaf node is a classification or decision. Figure 8-1 shows the structure of a decision tree.

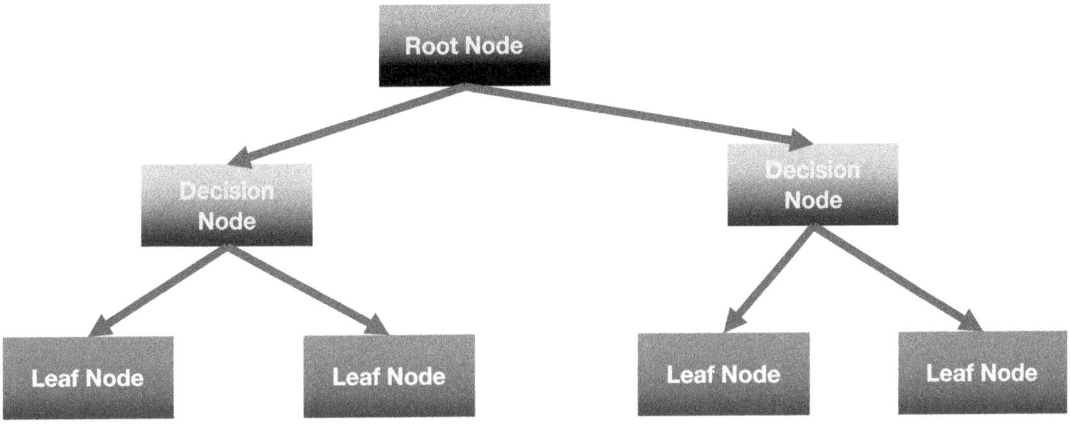

Figure 8-1. *Decision tree structure*

Develop the Decision Tree Classifier

Listing 8-1 develops a decision tree classifier with default hyperparameters.

Listing 8-1. Develop the Decision Tree Classifier with Default Hyperparameters

```
from sklearn.tree import plot_tree
from sklearn.tree import DecisionTreeClassifier
dt = DecisionTreeClassifier()
dt.fit(x_train,y_train)
```

Decision Tree Hyperparameter Optimization

There are dozens of tunable decision tree hyperparameters. We cover only two key hyperparameters: `criterion` and `max_depth`. Criteria include `gini` impurity and `entropy`. By default, the criterion is `gini`, which gives the probability of each label or class. A higher value shows high homogeneity. Listing 8-2 finds the optimal hyperparameters.

Listing 8-2. Hyperparameter Optimization

```
param_griddt = {"criterion":("gini","entropy"),
                "max_depth":[1,3,5,7,9,12]}
grid_modeldt = GridSearchCV(estimator=dt, param_grid=param_griddt)
grid_modeldt.fit(x_train,y_train)
print("Best score: ", grid_modeldt.best_score_, "Best parameters:",
grid_modeldt.best_params_)
```

Best score: 0.7182993469278954 Best parameters: {'criterion': 'entropy', 'max_depth': 3}

The previous results indicate that we should use `entropy` as the criterion and set 3 as the maximum depth of the tree. See Listing 8-3.

Listing 8-3. Finalize the Decision Tree Classifier

```
dt = DecisionTreeClassifier(criterion= 'entropy', max_depth= 3)
dt.fit(x_train,y_train)
```

Visualize Decision Trees

Listing 8-4 visualizes the decision tree (see Figure 8-2).

Listing 8-4. Visualize the Decision Tree

```
plt.figure(figsize=(24,14))
tree.plot_tree(dt, filled=True)
plt.show()
```

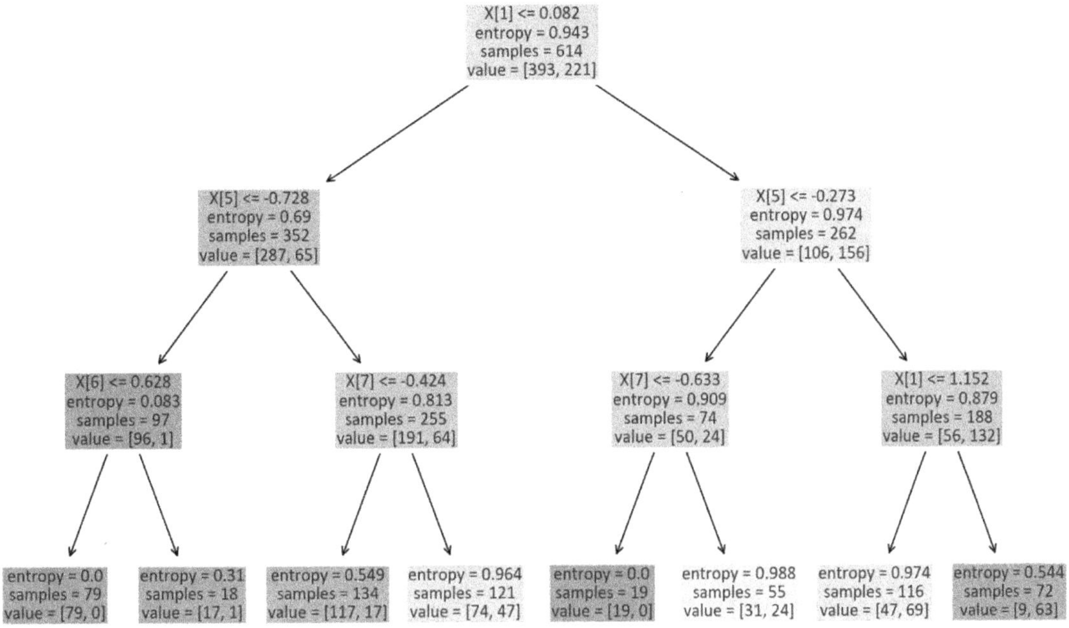

Figure 8-2. *Decision tree*

Feature Importance

Feature importance determines the relative importance of independent variables to a dependent variable. It allocates a score that ranges from 0 to 1 to each independent variable. An independent variable with the highest score is the most important variable, an independent variable with the second highest score is the second most important variable, and so forth. Listing 8-5 plots feature importance (see Figure 8-3).

Listing 8-5. Feature Importance

```
diabetes_features = [x for i,x in enumerate(df.columns) if i!=8]
def plot_feature_importances_diabetes(model):
    plt.figure(figsize=(8,6))
    n_features = 8
    plt.barh(range(n_features), model.feature_importances_, align='center')
    plt.yticks(np.arange(n_features), diabetes_features)
    plt.xlabel("Feature Importance")
    plt.ylabel("Feature")
    plt.ylim(-1, n_features)
```

```
plot_feature_importances_diabetes(dt)
plt.savefig('feature_importance')
```

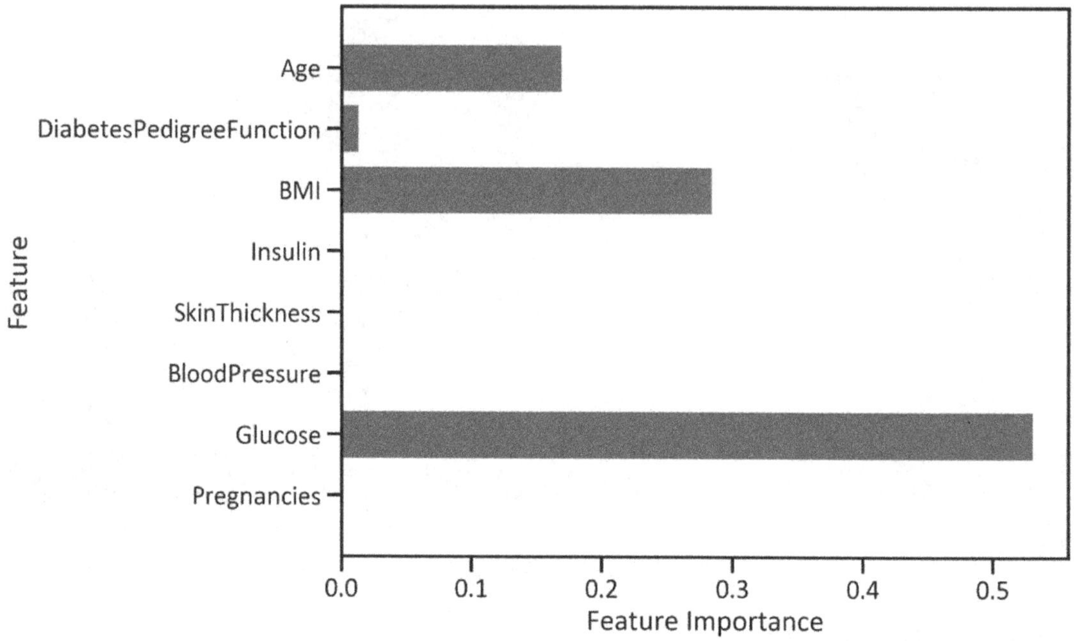

Figure 8-3. *Feature importance*

Figure 8-3 shows that there are a few important variables to the dependent variable. Glucose has the topmost score, meaning that it is more important than other variables to the dependent variable (diabetes outcome). This is followed by `bmi` and `age`, respectively.

We use the previous findings to reduce the number of variables in the data. Listing 8-6 reduces the data.

Listing 8-6. Re-processing

```
x = df[["Glucose","DiabetesPedigreeFunction","Age"]]
x = df.iloc[::,0:8]
y = df.iloc[::,-1]
x_train, x_test, y_train, y_test = train_test_split(x,y,test_size=0.2,
random_state=0)
scaler = StandardScaler()
x_train = scaler.fit_transform(x_train)
```

```
x_test = scaler.transform(x_test)
dt = DecisionTreeClassifier(criterion= 'entropy', max_depth= 3)
dt.fit(x_train,y_train)
```

Evaluate the Decision Tree Classifier

Listing 8-7 returns a table that compares actual values and predicted values (see Table 8-1).

Listing 8-7. Actual Values and Predicted Values

```
y_preddt = dt.predict(x_test)
pd.DataFrame({"Actual":y_test, "Predicted": y_preddt})
```

Table 8-1. *Actual Values and Predicted Values*

	Actual	Predicted
661	1	1
122	0	0
113	0	0
14	1	0
529	0	0
...	...	...
476	1	0
482	0	0
230	1	1
527	0	0
380	0	0

Confusion Matrix

Listing 8-8 and Table 8-2 show the performance of the decision tree classifier. It has four combinations of actual values and predicted values.

Listing 8-8. Confusion Matrix

```
cmatdt = pd.DataFrame(metrics.confusion_matrix(y_test,y_preddt),
index=["Actual: No","Actual: Yes"],
columns=("Predicted: No","Predicted: Yes"))
cmatdt
```

Table 8-2. *Confusion Matrix*

	Predicted: No	Predicted: Yes
Actual: No	93	14
Actual: Yes	23	24

Classification Report

To gain deep insight into the decision tree classifier's performance, we use the classification report. Listing 8-9 tabulates a classification report (see Table 8-3).

Listing 8-9. Classification Report

```
creportdt = pd.DataFrame(metrics.classification_report(y_test,y_preddt,
output_dict=True)).transpose()
creportdt
```

Table 8-3. *Classification Report*

	precision	recall	f1-score	support
0	0.801724	0.869159	0.834081	107.00000
1	0.631579	0.510638	0.564706	47.00000
accuracy	0.759740	0.759740	0.759740	0.75974
macro avg	0.716652	0.689899	0.699393	154.00000
weighted avg	0.749797	0.759740	0.751869	154.00000

Table 8-3 indicates that the decision tree classifier performs poorly. The accuracy score is 74%, which is about 8% lower than the accuracy score of the logistic regression classifier, linear support vector classifier, and linear discriminant analysis classifier. The classifier also has a lower precision score than both classifiers.

ROC Curve

The most reliable way involves predicting the probabilities of each class. First, we predict the probabilities of each class. Thereafter, we construct an ROC and find the AUC. Listing 8-10 produces an ROC curve (see Figure 8-4).

Listing 8-10. ROC Curve

```
y_preddt_proba = dt.predict_proba(x_test)[::,1]
fprdt, tprdt, _ = metrics.roc_curve(y_test,y_preddt_proba)
aucdt = metrics.roc_auc_score(y_test, y_preddt_proba)
plt.plot(fprdt, tprdt, label="AUC: "+str(aucdt), color="navy")
plt.plot([0,1],[0,1],color="red")
plt.xlim([0.00,1.01])
plt.ylim([0.00,1.01])
plt.xlabel("Specificity")
plt.ylabel("Sensitivity")
plt.legend(loc=4)
plt.show()
```

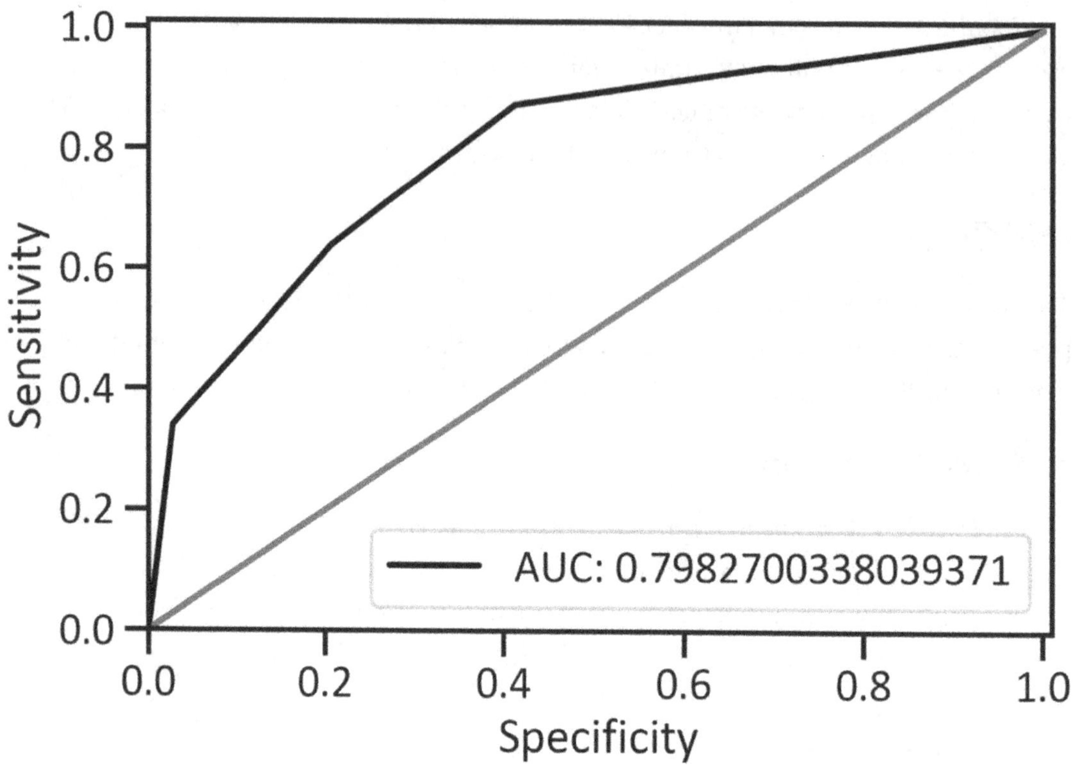

Figure 8-4. *ROC curve*

Figure 8-4 shows that the AUC score is close to 80%. Ideally, we want an AUC score greater than 80%. The previous AUC score shows that the classifier is not skillful in distinguishing between classes.

Precision-Recall Curve

Listing 8-11 produces a curve that gives meaningful insight about false positives and false negatives (see Figure 8-5).

Listing 8-11. Precision-Recall Curve

```
precisiondt, recalldt, thresholddt = metrics.precision_recall_curve(
y_test,y_preddt)
apsdt = metrics.roc_auc_score(y_test,y_preddt)
plt.plot(precisiondt, recalldt, label="APS: "+str(apsdt),color="navy")
plt.axhline(y=0.5,color="red")
plt.xlabel("Recall")
```

```
plt.ylabel("Precision")
plt.legend(loc=4)
plt.show()
```

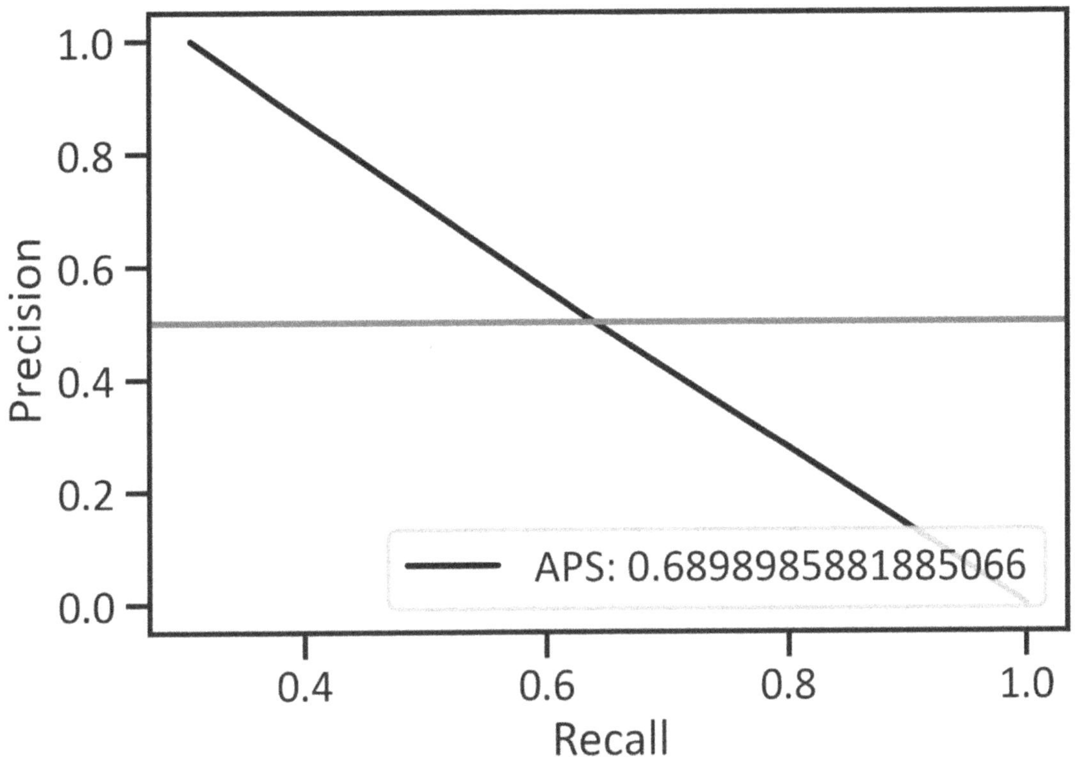

Figure 8-5. *Precision-recall curve*

Without considering the decision thresholds, the APS is 69%.

Learning Curve

Listing 8-12 produces a curve that depicts the progression of the LinearSVC classifier's accuracy as it learns the training data (see Figure 8-6).

Listing 8-12. Learning Curve

```
trainsizedt, trainscoredt, testscoredt = learning_curve(dt, x, y, cv=5,
n_jobs=5, train_sizes=np.linspace(0.1,1.0,50))
trainscoredt_mean = np.mean(trainscoredt,axis=1)
testscoredt_mean = np.mean(testscoredt,axis=1)
```

```
plt.plot(trainsizedt,trainscoredt_mean,color="red", label="Training Score")
plt.plot(trainsizedt,testscoredt_mean,color="navy", label="Cross Validation
Score")
plt.xlabel("Training Set Size")
plt.ylabel("Accuracy")
plt.legend(loc=4)
plt.show()
```

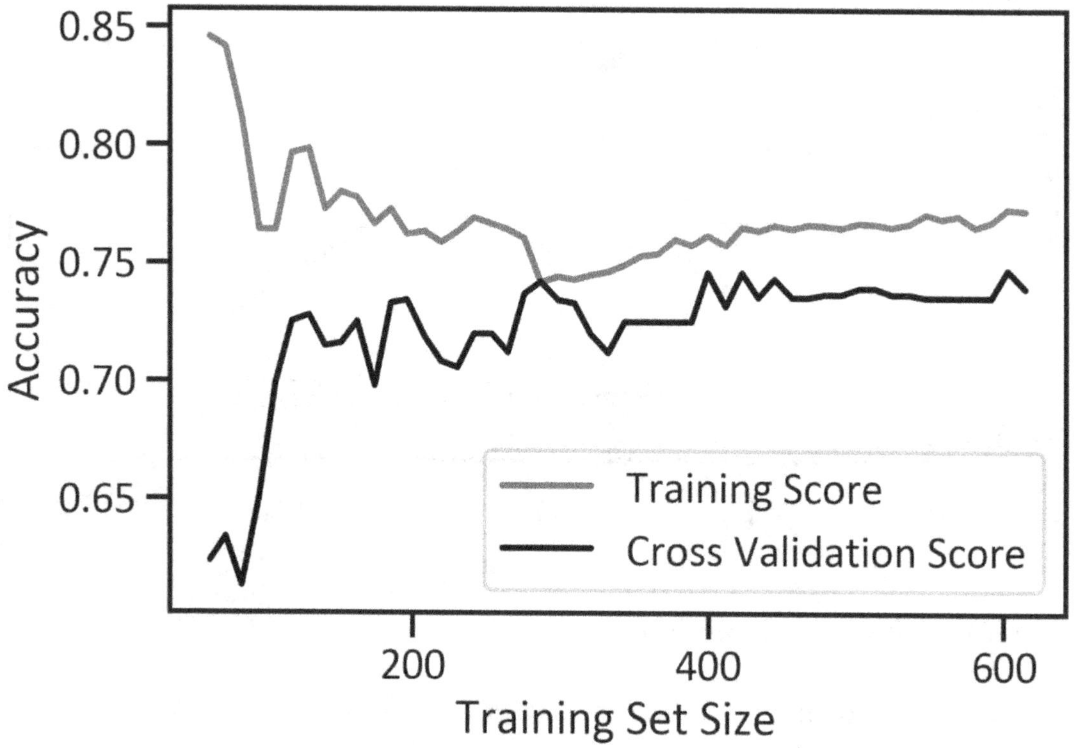

Figure 8-6. *Learning curve*

Figure 8-6 shows that the training accuracy score is higher than the cross validation from the time the decision tree classifier learns the data. The classifier starts off very optimistic, but as we increase the training set, the size drops and becomes comfortable around the 0.75 to 0.80 range.

Conclusion

This chapter showed the tree classification modeling technique, including discovering the optimal hyperparameters, finding variables that are the most important to the dependent variables, and visualizing the decision tree and classifier using only import variables and the best hyperparameters. There are alternative classifiers that uniquely extend the decision tree classifier, such as the extra tree and random forest tree. Extra trees split a point of the decision tree at random. A random forest tree iteratively builds multiple decision trees in parallel.

CHAPTER 9

Back to the Classics

We will now cover one of the oldest classification methods; naïve Bayes is an early 18th century model. It is a supervised learning model that solves binary and multiclass classification problems. The word *naïve* derives from the assumption that the model makes about the data. We consider it naïve because it assumes that variables are independent of each other, meaning there is no dependency on the data. This rarely occurs in the actual world. We can reduce the naïve Bayes theorem into Equation 9-1.

$$P(A|B) = (P(B|A)P(A))/(P(B)) \quad \text{(Equation 9-1)}$$

Here, the probability of event A happening knowing that event B has already happened.

This chapter covers Gaussian naïve Bayes. It assumes that the independent variables are binary. However, we can relax the assumption to accommodate continuous variables that follow a normal distribution. Also, it is memory efficient and fast because we do not perform hyperparameter tuning; it has no tunable hyperparameters. Bayes models are part of the probabilistic family. To comprehend the model, familiarize yourself with naïve Bayes.

The Naïve Bayes Theorem

When dealing with continuous independent variables, use the GaussianNB classifier. It is relatively easy to model, provided you have background on dealing with continuous variables. Refer to the first chapter of the book for a recap on the Gaussian distribution (also known as *normal distribution*). We mentioned that the data follows a normal distribution when the data points are close to the actual mean. The GaussianNB classifier estimates the probability of a variable belonging to a certain class; simultaneously, it estimates the central values, in other words, the mean value

© Tshepo Chris Nokeri 2021
T. C. Nokeri, *Data Science Revealed*, https://doi.org/10.1007/978-1-4842-6870-4_9

or standard deviation value. It is most suitable for features with high dimensionality. At most, we use this classifier when the training set is small. The classifier uses the Gaussian probability density function to estimate the probabilities of new values of the independent variable.

Develop the Gaussian Naïve Bayes Classifier

The GaussianNB classifier does not require hyperparameter optimization. Likewise, we obtained the data from Kaggle.[1] Listing 9-1 completes the GaussianNB classifier.

Listing 9-1. Finalize the GaussianNB Classifier

```
from sklearn.naive_bayes import GaussianNB
gnb = GaussianNB()
gnb.fit(x_train,y_train)
```

Listing 9-2 returns a table that highlights both actual classes and predicted classes (see Table 9-1).

Listing 9-2. Actual Values and Predicted Values

```
y_predgnb = gnb.predict(x_test)
pd.DataFrame({"Actual":y_test, "Predicted":y_predgnb})
```

[1] https://www.kaggle.com/uciml/pima-indians-diabetes-database

Table 9-1. Actual Values and Predicted Values

	Actual	Predicted
661	1	1
122	0	0
113	0	0
14	1	1
529	0	0
...	...	...
476	1	0
482	0	0
230	1	0
527	0	0
380	0	0

Evaluate the Gaussian Naïve Bayes Classifier

To understand how the GaussianNB classifier performs, we must compare actual classes and predicted classes side by side.

Confusion Matrix

Listing 9-3 and Table 9-2 produces a matrix that highlights the values that we used to estimate the key performance of the GaussianNB classifier.

Listing 9-3. Confusion Matrix

```
cmatgnb = pd.DataFrame(metrics.confusion_matrix(y_test,y_predgnb),
index=["Actual: No","Actual: Yes"],
                            columns=("Predicted: No","Predicted: Yes"))
cmatgnb
```

Table 9-2. *Confusion Matrix*

	Predicted: No	Predicted: Yes
Actual: No	93	14
Actual: Yes	18	29

Classification Report

Listing 9-4 and Table 9-3 highlight key classification evaluation metrics such as accuracy, precision, recall, and others.

Listing 9-4. Classification Report

```
creportgnb = pd.DataFrame(metrics.classification_report(y_test,y_predgnb,
output_dict=True)).transpose()
creportgnb
```

Table 9-3. *Classification Report*

	precision	recall	f1-score	Support
0	0.837838	0.869159	0.853211	107.000000
1	0.674419	0.617021	0.644444	47.000000
accuracy	0.792208	0.792208	0.792208	0.792208
macro avg	0.756128	0.743090	0.748828	154.000000
weighted avg	0.787963	0.792208	0.789497	154.000000

Table 9-3 shows that class 0 has the highest precision score (0.84) and recall score (0.87). Overall, the model is accurate 79% of the time.

ROC Curve

Listing 9-5 and Figure 9-1 represent different thresholds that enable an operator to trade off precision and recall. The threshold is between 0 and 1.

Listing 9-5. ROC Curve

```
y_predgnb_proba = gnb.predict_proba(x_test)[::,1]
fprgnb, tprgnb, _ = metrics.roc_curve(y_test,y_predgnb_proba)
aucgnb = metrics.roc_auc_score(y_test, y_predgnb_proba)
plt.plot(fprgnb, tprgnb, label="auc: "+str(aucgnb), color="navy")
plt.plot([0,1],[0,1],color="red")
plt.xlim([0.00,1.01])
plt.ylim([0.00,1.01])
plt.xlabel("Specificty")
plt.ylabel("Sensitivity")
plt.legend(loc=4)
plt.show()
```

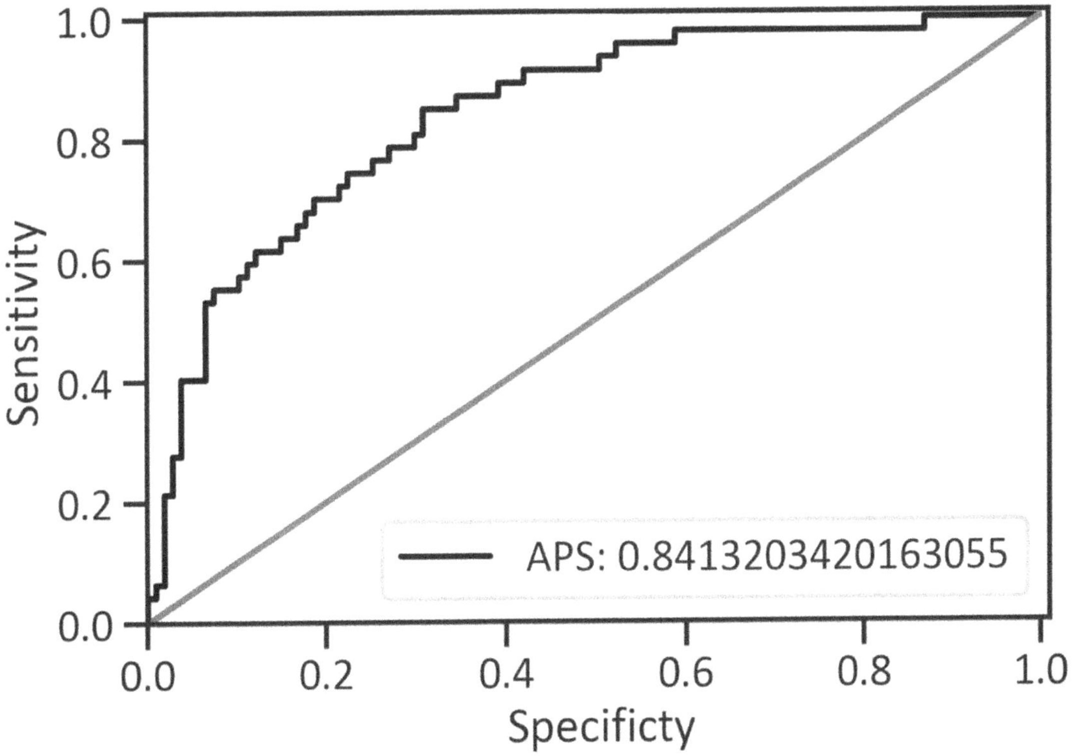

Figure 9-1. *ROC curve*

Figure 9-1 shows a curve that sails close to the right-side border when specificity is between 0 and 0.1. However, as specificity increases, the curve bends and slowly approaches 1. The AUC score is 0.84.

Precision Recall Curve

The classification report tells us that class 0 has an enormous number of data points (the data is imbalanced). We must focus more on precision and recall. Listing 9-6 produces the precision-recall curve (see Figure 9-2).

Listing 9-6. Precision-Recall Curve

```
precisiongnb, recallgnb, thresholdgnb = metrics.precision_recall_curve(
y_test,y_predgnb)
apsgnb = metrics.roc_auc_score(y_test,y_predgnb)
plt.plot(precisiongnb, recallgnb, label="aps: "+str(apsgnb),color="navy",
alpha=0.8)
plt.axhline(y=0.5,color="red",alpha=0.8)
plt.xlabel("Precision")
plt.ylabel("Recall")
plt.legend(loc=4)
plt.show()
```

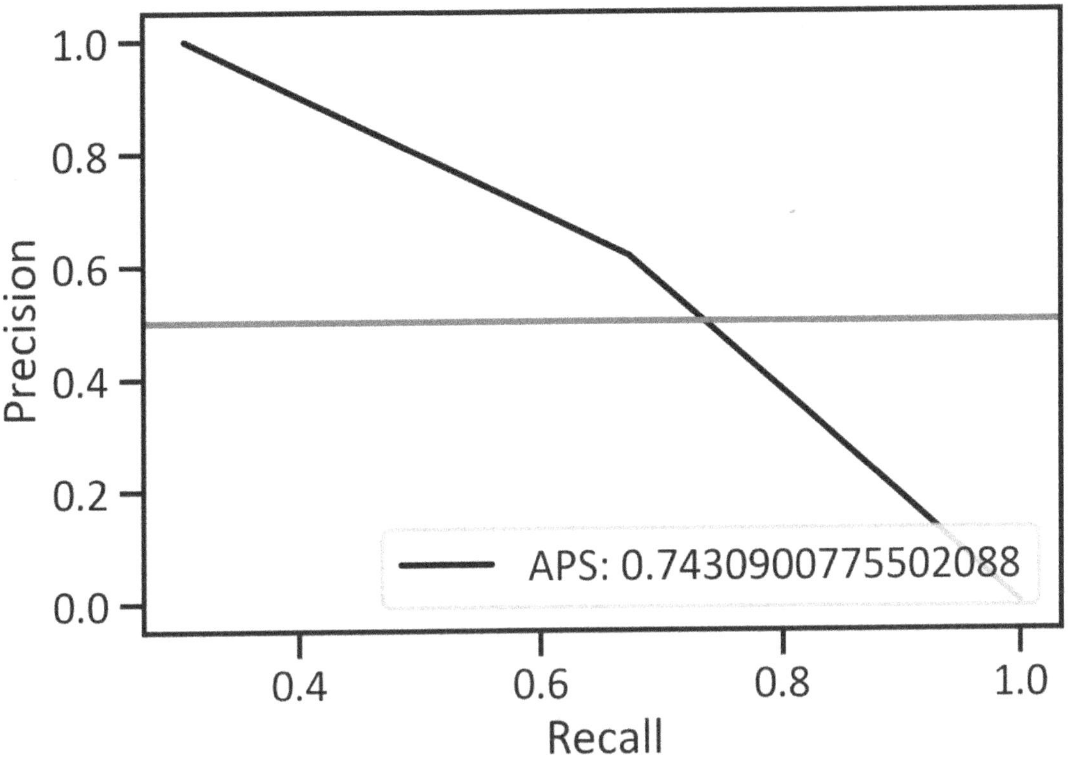

Figure 9-2. *Precision-recall curve*

The mean of the precision scores is 74%. The curve does not approach the uppermost border. Rather, it slows down smoothly until it reaches 0.8, and then the momentum declines.

Learning Curve

Listing 9-7 produces a curve that depicts the progression of the GaussianNB classifier's accuracy as it learns the training data (see Figure 9-3).

Listing 9-7. Learning Curve

```
trainsizegnb, trainscoregnb, testscoregnb = learning_curve(gnb, x, y, cv=5,
n_jobs=5, train_sizes=np.linspace(0.1,1.0,50))
trainscoregnb_mean = np.mean(trainscoregnb,axis=1)
testscoregnb_mean = np.mean(testscoregnb,axis=1)
```

```
plt.plot(trainsizegnb,trainscoregnb_mean,color="red", label="Training
Score", alpha=0.8)
plt.plot(trainsizegnb,testscoregnb_mean,color="navy", label="Cross
Validation Score", alpha=0.8)
plt.xlabel("Training Set Size")
plt.ylabel("Accuracy")
plt.legend(loc=4)
plt.show()
```

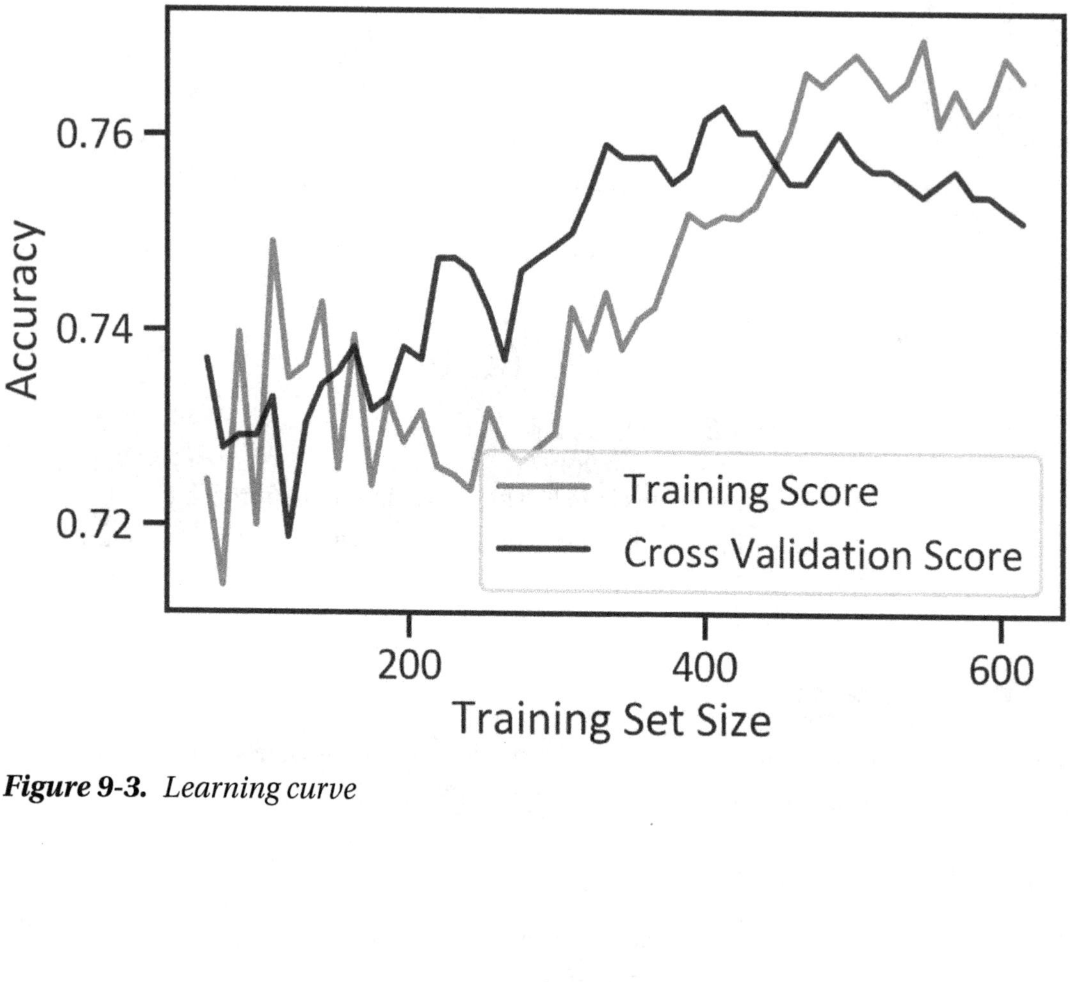

Figure 9-3. *Learning curve*

Figure 9-3 shows that the GaussianNB classifier is a fast learner. This comes as no surprise given that the model has no tunable hyperparameters. Compared to the classifiers we developed in the preceding chapters, the GaussianNB over-fit less from the beginning phase of the training process. Unlike in other learning curves, the training accuracy score does not crash in the first few training sets. Although the classifier is a fast learner, it failed to reach an accuracy score of 80%.

Conclusion

This chapter covered the naïve Bayes classifier. Remember, there are other naïve classifiers such as the multinomial naïve Bayes model and the Bernoulli naïve Bayes model.[2] Their use depends on the context.

[2] https://scikit-learn.org/stable/modules/naive_bayes.html

CHAPTER 10

Cluster Analysis

This chapter briefly covers the cluster analysis concept in a structured format. In previous chapters, we sufficiently covered supervised learning. In supervised learning, we present a model with a set of correct answers, and then we permit it to predict unseen data. Now, let's turn our attention a little. Imagine we have data with a set of variables and there is no independent variable of concern. In such a situation, we do not develop any plausible assumptions about a phenomenon.

This chapter introduces unsupervised learning. We do not present a model with a set of correct answers; rather, we allow it to make intelligent guesstimates. Unsupervised learning encompasses dimension reduction and cluster analysis. In this chapter, we first familiarize you with a dimension reduction method widely called principal component analysis (PCA), and we then acquaint you with cluster analysis. Thereafter, we develop and appraise models such as the K-means model, agglomerative model, and DBSCAN model. Before we preprocess and model the data, let's discuss cluster analysis.

What Is Cluster Analysis?

Cluster analysis is a method that assuredly finds a group of data points with similarities and dissimilarities. In cluster analysis, we are concerned with distinct clusters that explain variation in the data. There is no real dependent variable. We treat all variables equally. Popular clustering techniques include *centroid clustering*, which is random and selects centroids of the data points into a group of specified clusters, i.e., K-means; *density clustering*, which groups data points based on density population, i.e., DBSCAN; and *distribution clustering*, which identifies the probability of data points belonging in a cluster based on some distribution, i.e., Gaussian mixture model, etc.

© Tshepo Chris Nokeri 2021
T. C. Nokeri, *Data Science Revealed*, https://doi.org/10.1007/978-1-4842-6870-4_10

Cluster Analysis in Practice

Table 10-1 displays the data for this use case. There are three variables in the data. There is no actual independent variable. All variables are of equal importance. We are concerned with the similarities and dissimilarities in the data. Likewise, we obtained the example data from Kaggle.[1]

Table 10-1. *Dataset*

	Age	Annual Income (k$)	Spending Score (1–100)
0	19	15	39
1	21	15	81
2	20	16	6
3	23	16	77
4	31	17	40

The Correlation Matrix

We begin by computing the association between variables in the data. When there is a set of continuous variables in the data, use the Pearson method to measure correlation. Listing 10-1 produces the Pearson correlation matrix (see Table 10-2).

Listing 10-1. Pearson Correlation Matrix

```
dfcorr = df.corr()
dfcorr
```

Table 10-2. *Pearson Correlation Matrix*

	Age	Annual Income (k$)	Spending Score (1–100)
Age	1.000000	-0.012398	-0.327227
Annual Income (k$)	-0.012398	1.000000	0.009903
Spending Score (1–100)	-0.327227	0.009903	1.000000

[1] `https://www.kaggle.com/kandij/mall-customers`

Table 10-2 expresses a fragile adverse association between age and annual income and a sturdy adverse association between age and spending score.

The Covariance Matrix

The principal purpose of computing the covariance is to develop the eigen matrix. Listing 10-2 and Table 10-3 summarize the variability between the variables.

Listing 10-2. Covariance Matrix

```
dfcov = df.cov()
dfcov
```

Table 10-3. *Covariance Matrix*

	Age	Annual Income (k$)	Spending Score (1-100)
Age	195.133166	-4.548744	-118.040201
Annual Income (k$)	-4.548744	689.835578	6.716583
Spending Score (1-100)	-118.040201	6.716583	666.854271

At most, we are uninterested in the covariance between variables; instead, we are interested in the correlation between them and the severity of the correlation.

The Eigen Matrix

To reliably find the severity of the correlation between variables, use the eigenvalues. An eigenvalue less than 0 indicates multicollinearity, between 10 and 100 indicates slight multicollinearity, and more than 100 indicates severe multicollinearity. Listing 10-3 computes and tabulates eigenvalues and eigenvectors (see Table 10-4).

Listing 10-3. Eigen Matrix

```
eigenvalues, eigenvectors = np.linalg.eig(dfcov)
eigenvalues = pd.DataFrame(eigenvalues)
eigenvectors = pd.DataFrame(eigenvectors)
eigen = pd.concat([eigenvalues,eigenvectors],axis=1)
```

```
eigen.index = df.columns
eigen.columns = ("Eigen values","Age","Annual Income (k$)","Spending Score
(1-100)")
eigen
```

Table 10-4. *Eigen Matrix*

	Eigen values	Age	Annual Income (k$)	Spending Score (1–100)
Age	167.228524	0.973210	0.188974	-0.130965
Annual Income (k$)	700.264355	0.005517	-0.588641	-0.808376
Spending Score (1–100)	684.330136	0.229854	-0.785997	0.573914

Table 10-4 displays eigenvectors and eigenvalues. There is severe multicollinearity. As we prominently mentioned in the preceding chapters, multicollinearity can adversely affect the conclusions.

Listing 10-4 applies the `StandardScaler()` method to scale the data in such a way that the mean value is 0 and the standard deviation is 1.

Listing 10-4. Normalize Data

```
from sklearn.preprocessing import StandardScaler
scaler = StandardScaler()
df = scaler.fit_transform(df)
```

Next, we perform PCA.

Principal Component Analysis

In PCA, we use the eigenvalues to find the source of variation in the data. PCA computes the cumulative proportion. Remember that we do not base PCA on a model. It finds data points that explain variation. Like factor analysis, we use PCA[2] to streamline the structure of a set of variables. It calculates principal components as the linear combinations of the variables. The primary aim of the method is to explain the total

[2]https://scikit-learn.org/stable/modules/generated/sklearn.decomposition.PCA.html

variance. It also condenses data into fewer components so that the components used signifies model estimates. Last, it eliminates the factors with great eigenvalues (extreme variability). Not only that, but it uses the Kaiser's criterion to discover the number of factors to preserve. Figure 10-1 simplifies the structure of a set of variables.

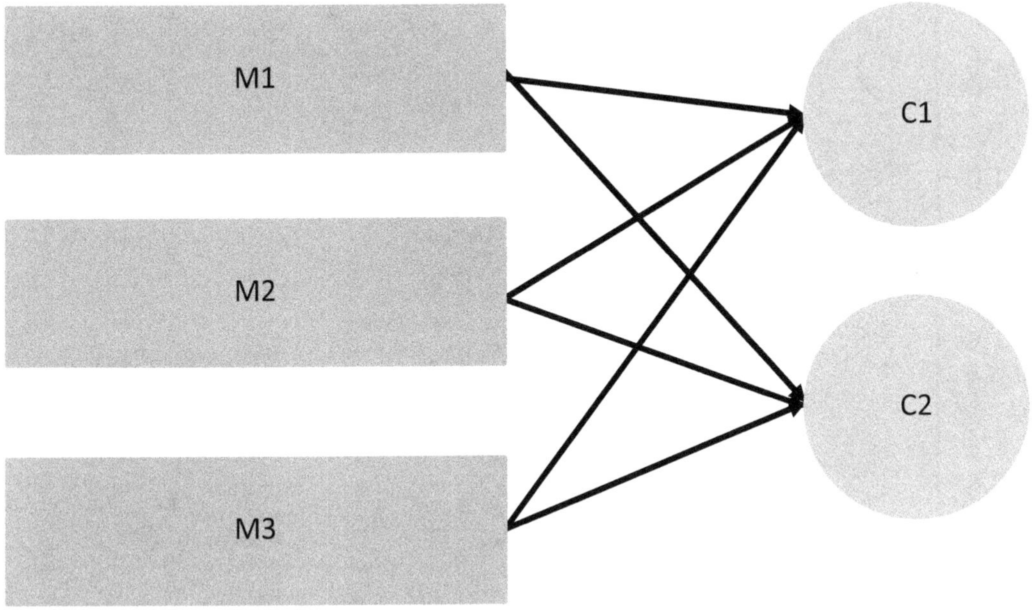

Figure 10-1. *Principal components analysis*

Here, the variables (M1, M2 and M3) are based on some factors or components, C1 and C2. Recall that we are interested in discovering variables that back principal components. Listing 10-5 returns PCA loading and creates a loading matrix.

Listing 10-5. Explained Variance

```
pca = PCA()
pca.fit_transform(df)
pca_variance = pca.explained_variance_
plt.figure(figsize=(8, 6))
plt.bar(range(3), pca_variance, align="center", label="Individual Variance")
plt.legend()
plt.ylabel("Variance Ratio")
plt.xlabel("Principal Components")
plt.show()
```

Figure 10-2 shows the ranking of the components' relative importance to the predictor variable. It confirms that the first component explains most of the variation in the data. See Listing 10-6.

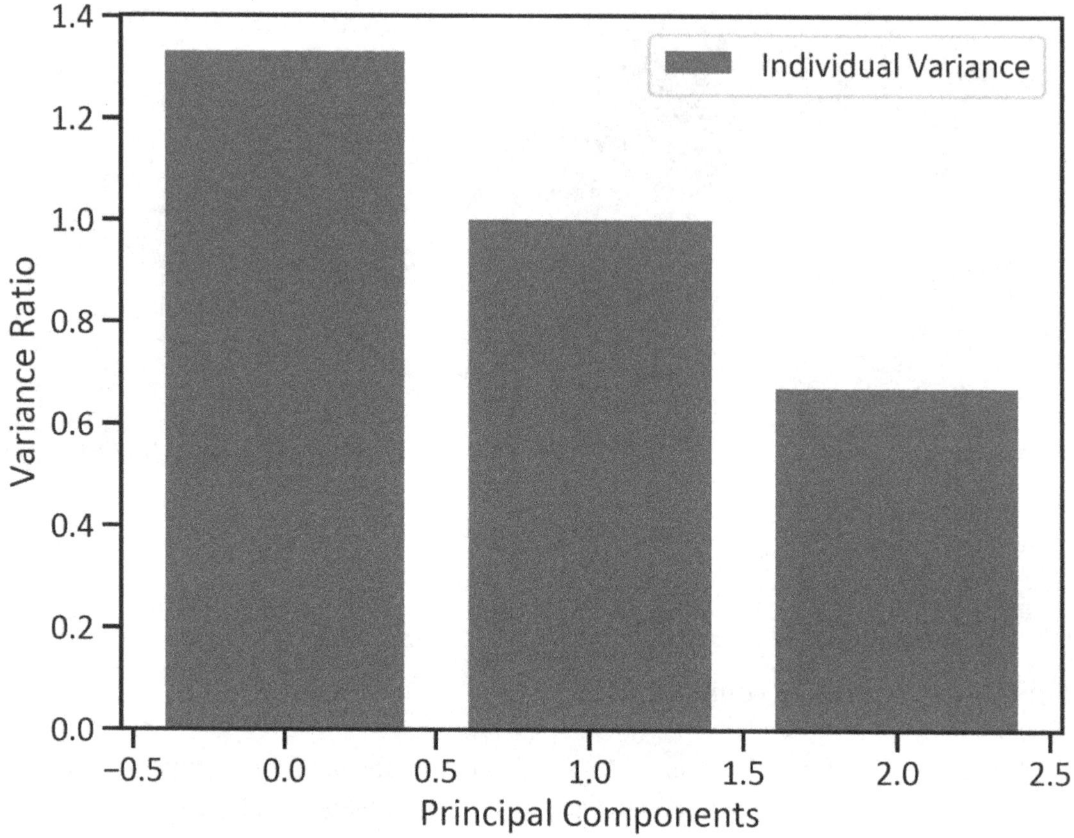

Figure 10-2. *Explained variance*

Listing 10-6. Principal Components Matrix

```
pca = PCA(n_components=3).fit(df)
pca_components = pca.components_.T
pca_components = pd.DataFrame(pca_components)
pca_components.index = df.columns
pca_components.columns = df.columns
pca_components
```

Listing 10-6 returns a conditional index that depicts PCA loading of each component (see Table 10-5).

Table 10-5. Principal Components

	Age	Annual Income (k$)	Spending Score (1–100)
Age	-0.188974	0.130965	0.973210
Annual Income (k$)	0.588641	0.808376	0.005517
Spending Score (1–100)	0.785997	-0.573914	0.229854

Age has the uppermost PCA loading (it is the first component), followed by annual income (k$) and so forth. Listing 10-7 transforms the data. Remember, we specify the number of components as 3.

Listing 10-7. Finalize PCA

```
pca2 = PCA(n_components=3)
pca2.fit(df)
x_3d = pca2.transform(df)
plt.figure(figsize=(8,6))
plt.scatter(x_3d[:,0], x_3d[:,2], c=old_df['Annual Income (k$)'])
plt.xlabel("y")
plt.show()
```

After performing dimension reduction, we plot the data points using a scatter plot. Figure 10-3 shows the distribution of the data points. See Listing 10-8.

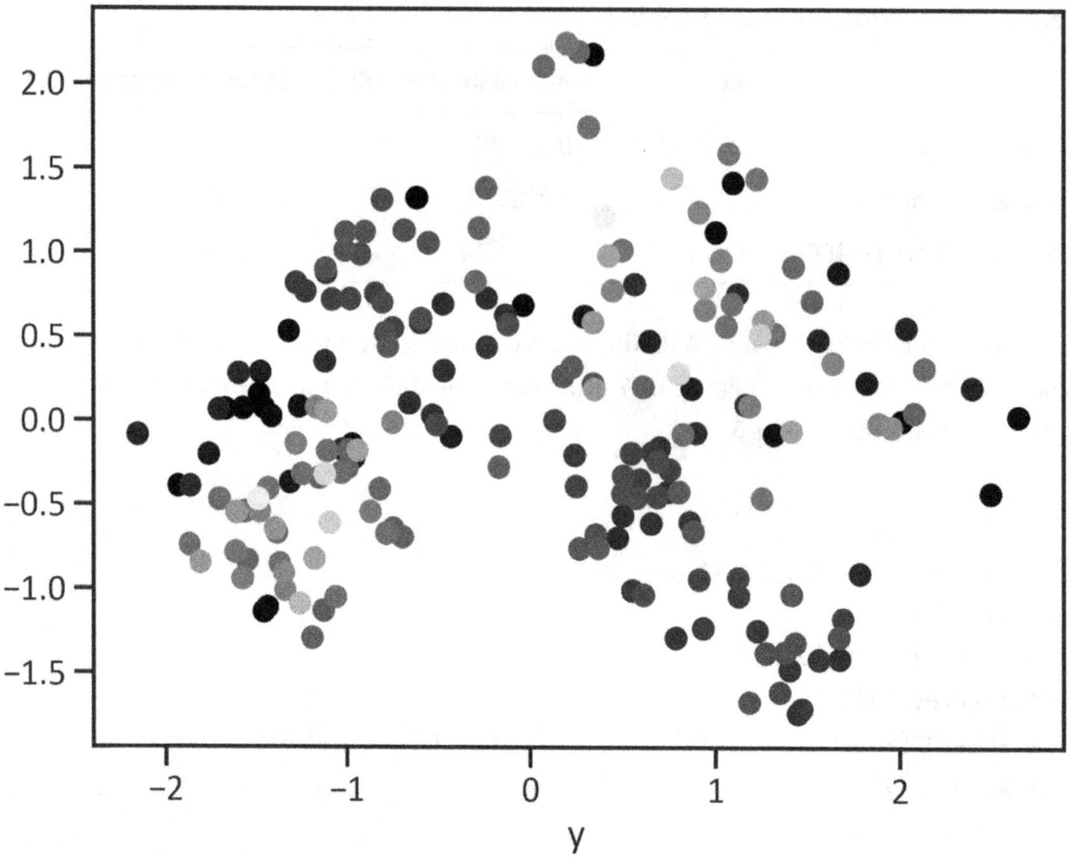

Figure 10-3. *Principal components*

Listing 10-8. Reduce Data Dimension

```
df = pca.transform(df)
```

Elbow Curve

An elbow curve is widely recognized as a *knee of curve* plot. We use it to discover the number of clusters in the data. It has the number of clusters on the x-axis and the percent of the variance explained on the y-axis. We use a cutoff point to select the number of clusters. A cutoff point is a point at which a smooth bend ends and a sharp bend begins. See Listing 10-9.

Listing 10-9. Elbow Curve

```
Nc = range(1,20)
kmeans = [KMeans(n_clusters=i) for i in Nc]
scores = [kmeans[i].fit(df).score(df) for i in range(len(kmeans))]
fig, ax = plt.subplots()
plt.plot(Nc, scores)
plt.xlabel("No. of clusters")
plt.show()
```

Figure 10-4 displays a smooth bend from cluster 1 to cluster 3. However, from cluster 3, the curve bends abruptly. We use 3 as the cutoff point.

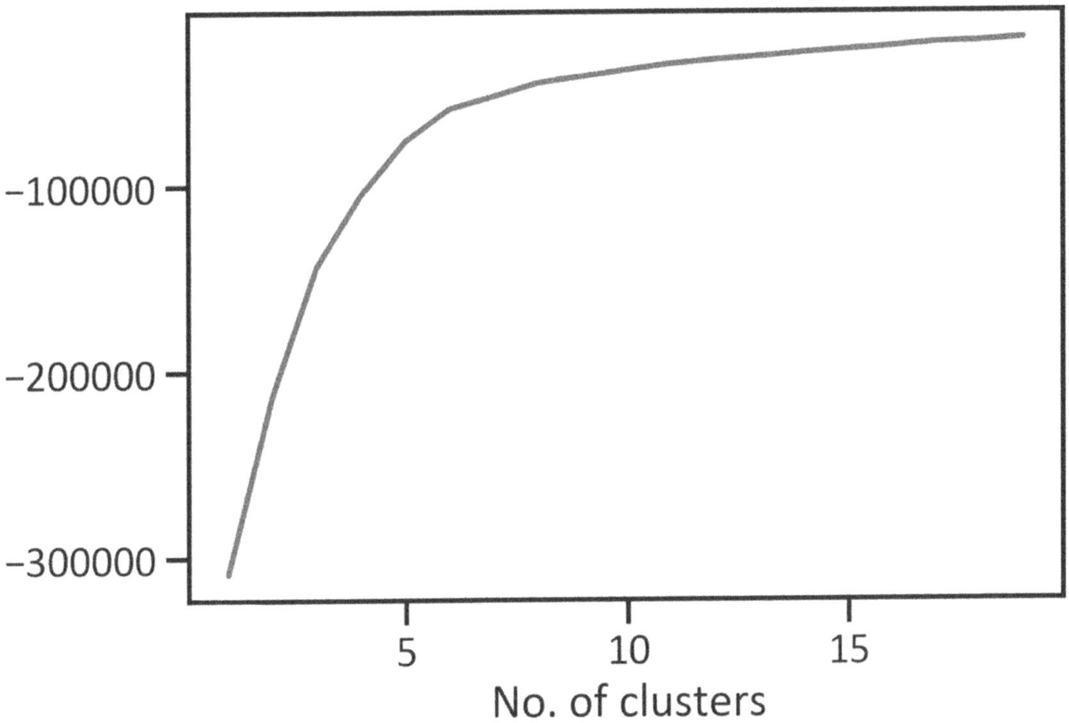

Figure 10-4.

There are various ways of skinning a cat. You can also use a scree plot to endorse the cutoff point. Remember that we normally use a scree plot in PCA to find the number of components to keep. Listing 10-10 produces a scree plot, which we can also use to determine the number of clusters to use for a cluster model (see Figure 10-5).

Listing 10-10. Scree Plot

```
ks = range(1,20)
ds = []
for k in ks:
    cls = KMeans(n_clusters=k)
    cls.fit(df)
    ds.append(cls.inertia_)
fig, ax = plt.subplots()
plt.plot(ks, ds)
plt.xlabel("Value of k")
plt.ylabel("Distortion")
plt.show()
```

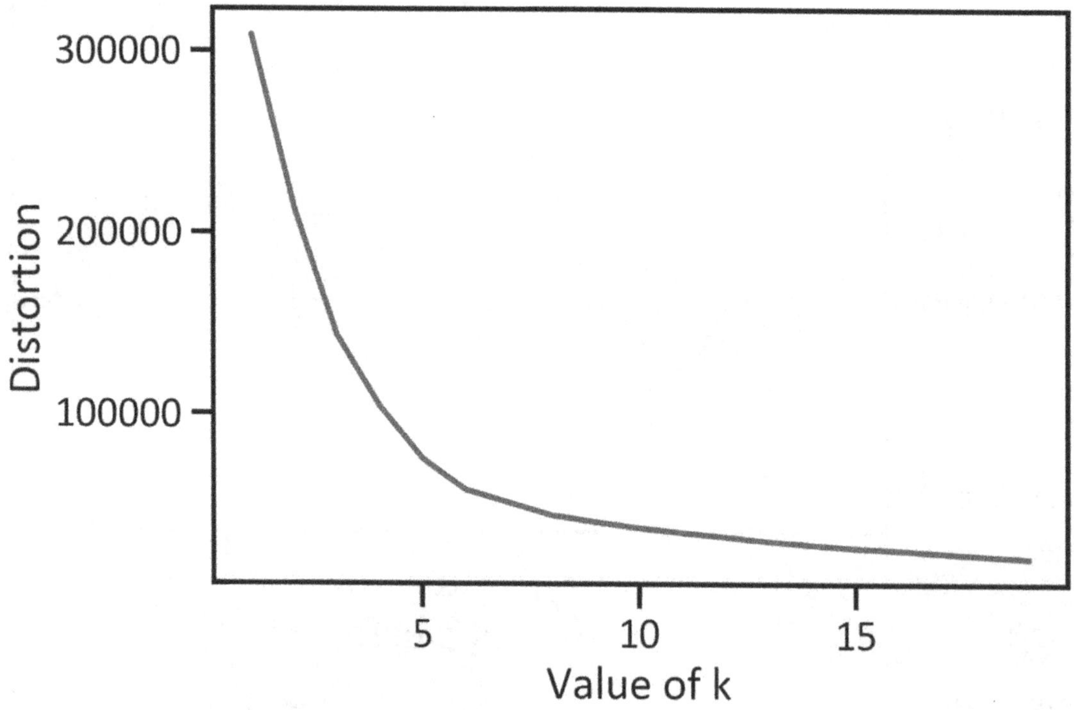

Figure 10-5. *Scree plot*

Figure 10-5 approves the scree plot point.

K-Means Clustering

The K-means model is the most prevalent cluster model. It splits the data into *k* clusters with the nearest mean *centroids*, and it then finds the distance between subsets to develop a cluster. It also reduces the intracluster distances and advance intercluster. We express the formula as shown in Equation 10-1.

$$Dis(x_1, x_2) = \sqrt{\sum_{i=0}^{n} (x_{1i} - y_{2i})^2} \qquad \text{(Equation 10-1)}$$

Basically, the model finds the initial *k* (the number of clusters) and calculates the distance between clusters. Thereafter, it allocates each data point to the closest centroid. The model works best with large samples. See Listing 10-11.

Listing 10-11. K-Means Model Using Default Hyperparameters

```
kmeans = KMeans(n_clusters=3)
kmeans_output = kmeans.fit(df)
kmeans_output
```

K-Means Hyperparameter Optimization

At most, we do not want to fit a model with initial hyperparameters, unless it yields optimal performance. With each use case, we must find the best hyperparameters (provided a model has tunable hyperparameters). Table 10-6 outlines key tunable K-means hyperparameters.

Table 10-6. *Tunable Hyperparameters*

Parameter	Description
`copy_X`	Finds whether the independent variable must be copied.
`tol`	Finds the precision. The default value is 0.003.
`max_iter`	Finds the maximum number of iterations.
`tol`	Finds the tolerance for the optimization

Listing 10-12 finds the optimal hyperparameters for the K-means model.

Listing 10-12. Hyperparameter Optimization

```
param_gridkmeans = {"copy_x":[False,True],
                    "max_iter":[1,10,100,1000],
                    "n_init":[5,10,15,20],
                    "tol":[0.0001,0.001,0.01,1.0]}
grid_modelkmeans = GridSearchCV(estimator=kmeans_output, param_grid=param_
gridkmeans)
grid_modelkmeans.fit(df)
print("Best score: ", grid_modelkmeans.best_score_, "Best hyper-parameters: ",
grid_modelkmeans.best_params_)
```

Best score: -49470.220198874384 Best hyper-parameters: {'copy_x': False, 'max_iter': 1, 'n_init': 15, 'tol': 0.01}

Listing 10-13 concludes the K-means model with optimal hyperparameters.

Listing 10-13. Finalize the K-Means Model

```
kmeans = KMeans(n_clusters=3,
                copy_x=False,
                max_iter= 1, n_init= 15,
                tol= 0.01)
kmeans_output = kmeans.fit(df)
kmeans_output
```

Listing 10-14 tabulates predicted labels (Table 10-7).

Listing 10-14. Predicted Labels

```
y_predkmeans = pd.DataFrame(kmeans_output.labels_, columns = ["Predicted"])
y_predkmeans
```

Table 10-7. *Predicted Labels*

	Predicted
0	1
1	1
2	1
3	1
4	1
...	...
195	0
196	2
197	0
198	2
199	0

Table 10-7 does not tell us much. We are basically observing labels produced by the K-means model.

Centroids

The central tendency of the data indicates much about the data. We must find the centers of clusters also recognized as centroids. Listing 10-15 applies `cluster_centers_` to find centroids (see Table 10-8).

Listing 10-15. Centroids

```
kmean_centroids = pd.DataFrame(kmeans_output.cluster_centers_,
columns = ("Cluster 1","Cluster 2","Cluster 3"))
kmean_centroids
```

Table 10-8. Centroids

	Cluster 1	Cluster 2	Cluster 3
0	-8.879001	-11.467679	0.891796
1	-10.435900	42.928296	-5.127039
2	43.326081	2.774284	1.470721

To really make sense of the model, Listing 10-16 creates a scatter plot displaying clusters and centers of cluster (see Figure 10-6).

Listing 10-16. Visualize the K-Means Model

```
fig, ax = plt.subplots()
plt.scatter(df[:,0],df[:,1],c=kmeans_output.labels_,cmap="viridis",s=20)
plt.scatter(kmean_centroids[:,0], kmean_centroids[:,1], color="red")
plt.xlabel("y")
plt.show()
```

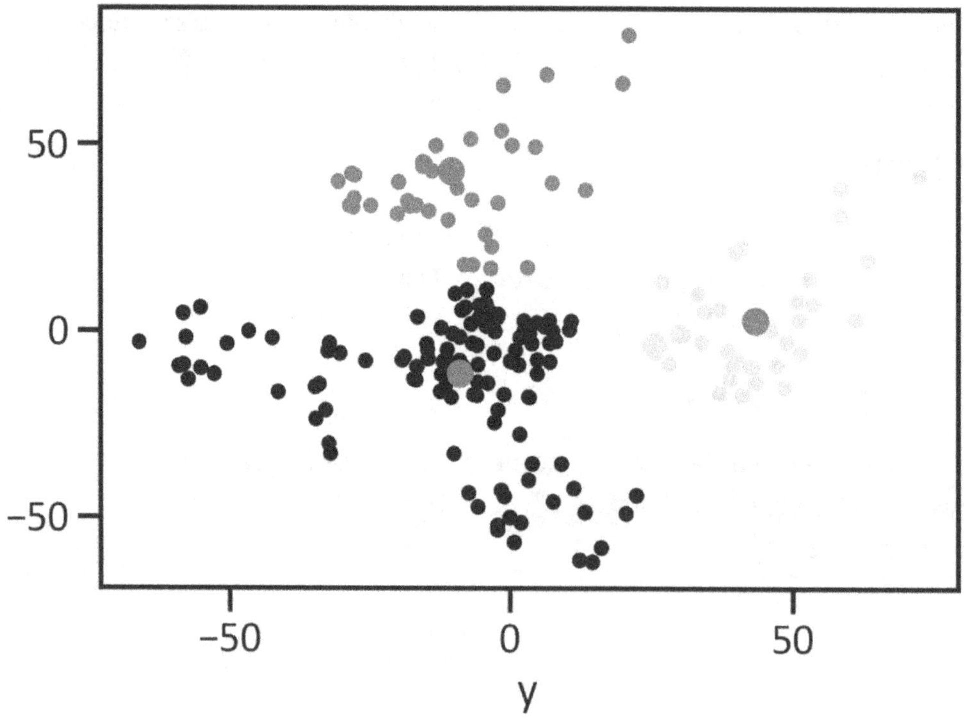

Figure 10-6. *K-means model*

The K-means model made intelligent guesstimates until data points were allocated to the nearest centroid and found the mean value of the centroids. Figure 10-6 indicates that there are three apparent clusters in the data.

The Silhouette Score

We use the Silhouette method to appraise the K-means model. The method compares cohesion and separation. It has values that range from -1 to 1, where -1 indicates severe mislabeling, 0 indicates overlaps in clusters, and 1 indicates that the sample is far away from adjacent clusters. See Listing 10-17.

Listing 10-17. Silhouette Score

```
metrics.silhouette_score(df, y_predkmeans)
0.3839349967742105
```

The Silhouette score is 0.38, which indicates that the model is far from perfect. There are overlaps in clusters.

Agglomerative Clustering

The agglomerative model is a hierarchical model that uses a bottom-up approach to produce dendrograms. It develops *n* clusters (one for each) and then computes proximity. Thereafter, it consolidates clusters close to each other while updating the proximity matrix. The end result is a single cluster. There are several ways of calculating the distance between data points. Table 10-9 highlights key distance calculation techniques.

Table 10-9. *Distancing Techniques*

Technique	Description
Single-linkage clustering	Finding the minimum distance between clusters
Complete-linkage clustering	Finding the maximum distance between clusters
Average linkage clustering	Finding the average distance between clusters
Centroid linkage clustering	Finding the distance between cluster centroids

Unlike the K-means model that randomly initializes centroids, the agglomerative generates more than one partition based on resolution. Listing 10-18 completes the agglomerative model.

Listing 10-18. Finalize the Agglomerative Model

```
agglo = AgglomerativeClustering(n_clusters=None,distance_threshold=0)
agglo_output = agglo.fit(pca_df)
agglo_output
Listing 10-19.
```

After finalizing the agglomerative model, Listing 10-19 finds and tabulates the predicted labels (Table 10-10). Listing 10-20 shows the agglomerative model.

Listing 10-19. Predicted Labels

```
y_pred = pd.DataFrame(agglo_output.labels_,columns = ["Predicted"])
y_pred
```

Table 10-10. *Predicted Labels*

	Predicted
0	-1
1	-1
2	-1
3	-1
4	-1
...	...
195	-1
196	-1
197	-1
198	-1
199	-1

Listing 10-20. Agglomerative Model

```
fig, ax = plt.subplots()
plt.scatter(df[:,0],df[:,1],c=agglo_output.labels_,cmap="viridis",s=20)
plt.xlabel("y")
plt.show()
```

Figure 10-7 points out three clusters. The large cluster breaks down into two small clusters. We also realize a minor overlap between cluster 1 and cluster 2.

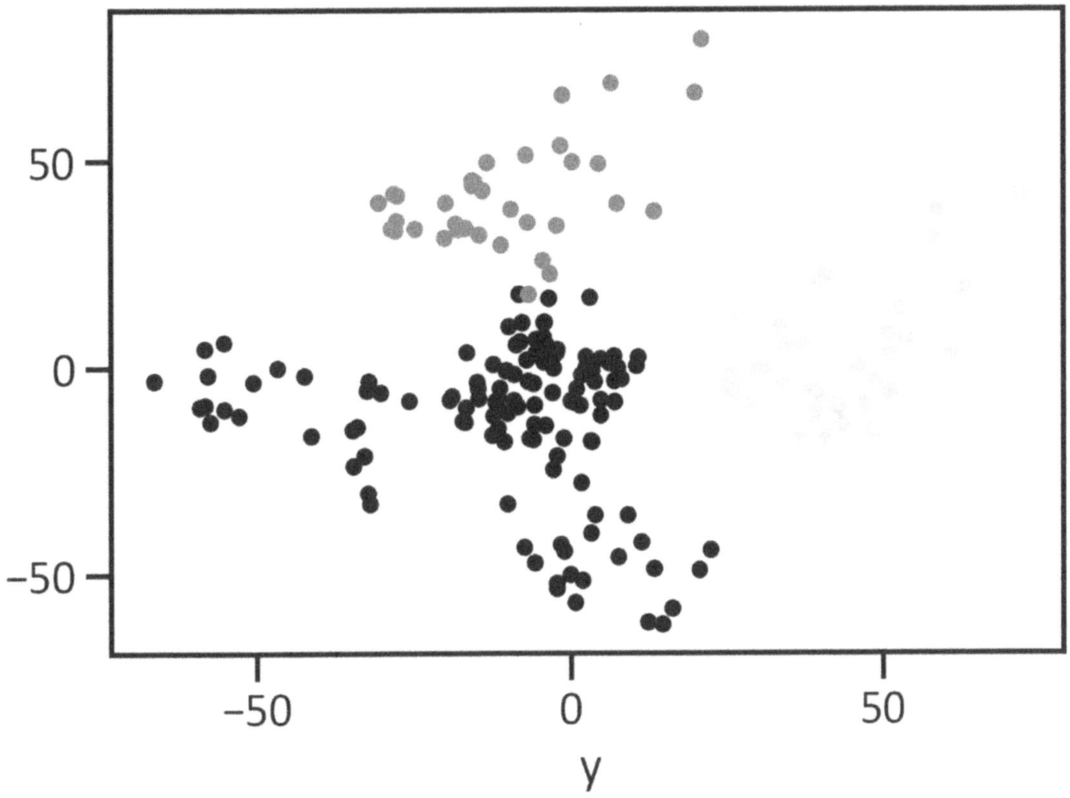

Figure 10-7. *Agglomerative model*

We use a dendogram to depict the hierarchy of the cluster tree. Listing 10-21 creates a linkage matrix, creates the counts of samples under each node, and creates a corresponding dendrogram (see Figure 10-8).

Listing 10-21. Visualize the Dendrogram

```
def plot_dendrogram(model, **kwargs):
    counts = np.zeros(model.children_.shape[0])
    n_samples = len(model.labels_)
    for i, merge in enumerate(model.children_):
        current_count = 0
        for child_idx in merge:
            if child_idx < n_samples:
                current_count += 1
            else:
                current_count += counts[child_idx - n_samples]
        counts[i] = current_count

    linkage_matrix = np.column_stack([model.children_, model.distances_,
                                      counts]).astype(float)
    dendrogram(linkage_matrix, **kwargs)
plot_dendrogram(agglo_output, truncate_mode='level', p=3)
```

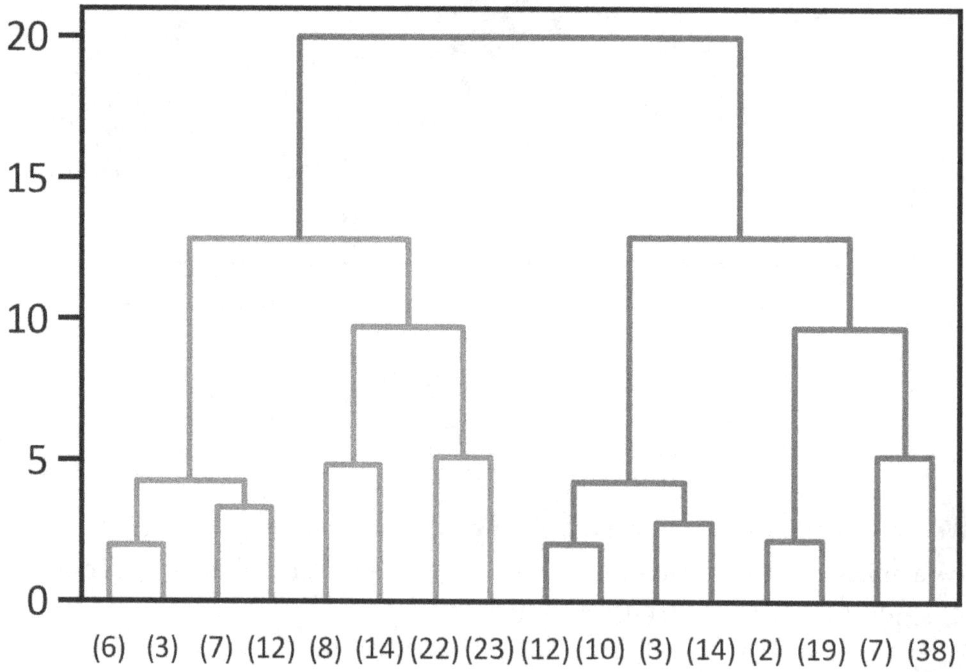

Figure 10-8. *Agglomerative dendrogram*

Figure 10-8 shows data points connected in clusters (one parent with two children).

Density-Based Spatial Clustering Model with Noise

We use the density-based spatial clustering model with noise (DBSCAN) model to find arbitrary-spatial clusters. It recognizes a radius area that includes several data points. Any of the data points can be a core point, border point, or outlier point. Points are grouped as clusters based on these points. Table 10-11 highlights these points.

Table 10-11. *DBSCAN Points*

Points	Description
Core point	A point enclosed by the smallest number of neighbors. Core points are surrounded by border points.
Border point	A point that is not enclosed by the smallest number of neighbors and can be reached from the core point.
Outlier point	A point that is neither a core point nor a border point. This is a point that is far away from the reach of a core point.

It finds core points that are neighbors and places them in the same cluster. As a result, a cluster encompasses at least one core point (reachable points) or border points. Moreover, it discovers clusters enclosed by different clusters. See Listing 10-22.

Listing 10-22. Finalize the DBSCAN Model

```
dbscan = DBSCAN()
dbscan_output = dbscan.fit(df)
dbscan_output
```

Listing 10-23 and Table 10-12 highlight predicted labels.

Listing 10-23. Predicted Labels

```
y_pred = pd.DataFrame(dbscan_output.labels_,columns = ["Predicted"])
y_pred
```

Table 10-12. Predicted Classes

	Predicted
0	-1
1	-1
2	-1
3	-1
4	-1
...	...
195	-1
196	-1
197	-1
198	-1
199	-1

Listing 10-24 characterizes the DBSCAN model (see Figure 10-9).

Listing 10-24. Visualize the DBSCAN Model

```
fig, ax = plt.subplots()
plt.scatter(df[:,0],df[:,1],c=dbscan_output.labels_,cmap="viridis",s=20)
plt.xlabel("y")
plt.show()
```

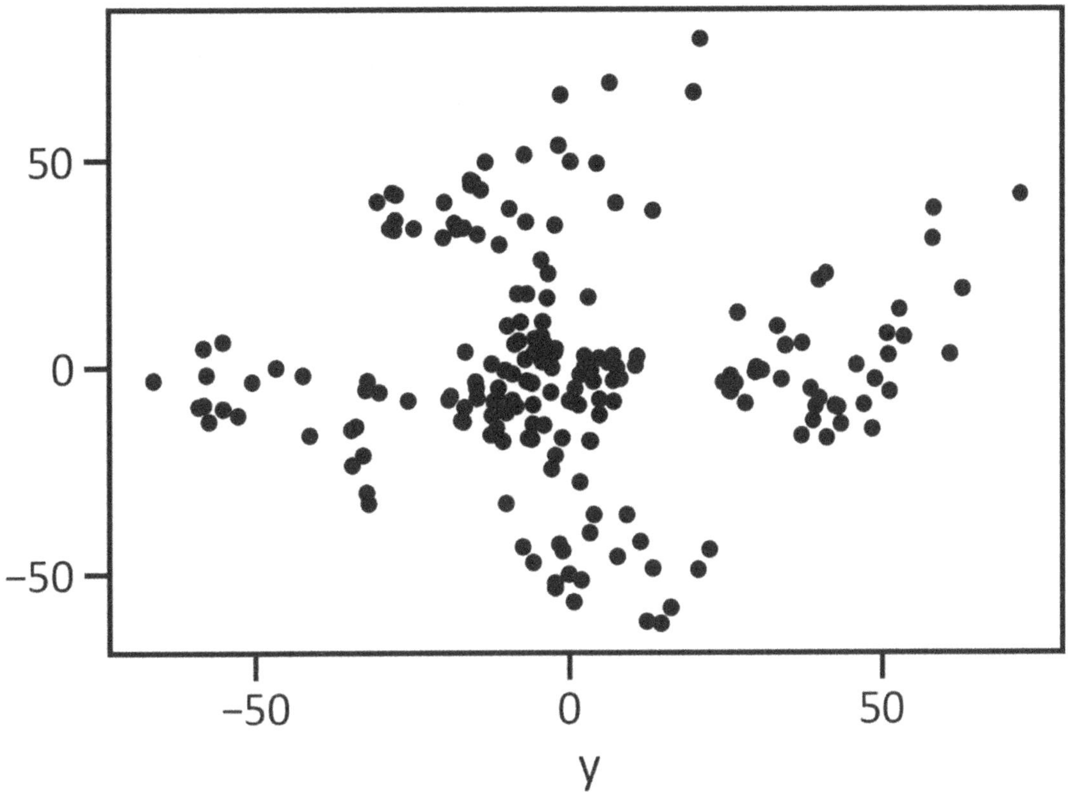

Figure 10-9. *DBSCAN model*

Figure 10-9 shows that the DBSCAN model does not make ample intelligent guesstimates. This does not entail that the model performs poorly in all settings. The model is suitable when the density-based criterion is met. It is apparent that Gaussian distributions demonstrate superior performance.

Conclusion

This chapter presented dimension reduction and cluster analysis. We developed and assessed three cluster models, namely, the K-means, DBSCAN model, and agglomerative model. We began by plummeting the dimension of the data using the PCA method. Thereafter, we employed the elbow curve to discover the number of clusters to be specified on the K-means model and agglomerative model. We do not postulate the number of clusters when developing the DBSCAN model.

Cluster models are relaxed; they are free of strong assumptions. The daunting part about cluster analysis is model evaluation. They lack robust evaluation metrics, and we rest on the Silhouette method for model evaluation.

CHAPTER 11

Survival Analysis

This chapter introduces the survival analysis. The method comprises two families: the parametric method and the nonparametric method. Parametric survival analysis models include Weibull, exponential, log-linear, and many others (learn more from Lifelines[1]). Nonparametric survival analysis models include Kaplan-Meier and Nelson-Aalen.

This chapter covers only the most popular nonparametric model, called the Kaplan-Meier model, which measures a binary variable using time. It has relaxed assumptions about the structure of the data. Likewise, it does not hold assumptions about the linearity and normality of the data.

It does not capture time in the traditional sense. It captures the actual date of entry and exit in the study. We also call survival analysis *reliability analysis* or *duration modeling*. When we use it, the duration is more important than time. In clinical trials, we use it to estimate the probabilities of a patient surviving an event (i.e., illness or death) beyond a specific time. In engineering, we use it to estimate the probabilities of a machine surviving an event (i.e., machine failure) beyond time.

It must fulfill certain conditions for us to consider a study and a survival analysis model reliable. First, the research design should include a control group. Second, there must be rigorous methods in place to capture the activities of participants in the study; we must always consider factors like a patient entering or leaving a study, a patient at risk, and a patient dying. Table 11-1 outlines key survival analysis terms.

[1] `https://lifelines.readthedocs.io/en/latest/Survival%20analysis%20with%20lifelines.html`

© Tshepo Chris Nokeri 2021
T. C. Nokeri, *Data Science Revealed*, https://doi.org/10.1007/978-1-4842-6870-4_11

Table 11-1. *Survival Analysis Key Term*

Term	Description
Survival time	Represents the timeline of events
Event	Represents a binary outcome under investigation (mostly illness or death)
Censoring	Represents removal of a patient from a study

Studies that measure event outcomes such as illness and death take a long period. Participants can enter or exit a study, some can die, and we report others as risky. In survival analysis, we mostly deal with missing values. We call data with missing values *censored data*. Unlike other nonparametric models such as logistic regression and linear discriminant analysis, survival analysis models are not sensitive to missing values. In fact, we always expect to find missing values in the data.

Survival Analysis in Practice

We loaded a dataset available in the Lifeline package, named `load_watons`. Lifeline is not a standard Python library. To install it in the Python environment, use `pip install lifelines`, and to install it in the Conda environment, use `conda install -c conda-forge lifelines`. The data is comprised of three columns. The first column is time (an independent variable), the second column is the event (a dependent variable), and the third column specifies the group the patient is in. The two groups are the control group and the miR-137 group. A patient can be in either category. See Listing 11-1 and Table 11-2.

Listing 11-1. Load Data

```
from lifelines.datasets import load_walton
import pandas as pd
df = load_waton()
df.head()
```

Table 11-2. Dataset

	T	E	Group
0	6.0	1	miR-137
1	13.0	1	miR-137
2	13.0	1	miR-137
3	13.0	1	miR-137
4	19.0	1	miR-137

Listing 11-2 returns the data structure.

Listing 11-2. Data Structure

```
tdf.info()
RangeIndex: 163 entries, 0 to 162
Data columns (total 3 columns):
 #   Column  Non-Null Count  Dtype
---  ------  --------------  -----
 0   T       163 non-null    float64
 1   E       163 non-null    int64
 2   group   163 non-null    object
dtypes: float64(1), int64(1), object(1)
memory usage: 3.9+ K
```

As you can see, 163 patients took part in the study. There are no missing values in the data. This means that the data collection methods used were rigorous. Remember, this hardly happens in the actual world. At most, we deal with data with missing values. The data is comprised of three columns; the first column is a continuous variable (timeline specified by the number of years), the second column is a binary variable (event outcome—we code a patient failing to survive beyond a specific time as 0 and a patient who survives beyond a specific time as 1), and the third column is a binary categorical (a patient is in the control group or the miR-137 group). Based on the structure of the data, we can use a survival analysis model such as the Kaplan-Meier model or the Nelson-Aalen model. In this chapter, we look at the Kaplan-Meier model. Before we develop the model, let's first describe the data.

Listing 11-3 counts and tabulates patients in each group (see Table 11-3).

Listing 11-3. Count of Groups

```
class_series_group = df.groupby("group").size()
class_series_group = pd.DataFrame(class_series_group, columns=["Count"])
class_series_group
```

Table 11-3. *Count of Patients in Each Group*

	Count
group	
control	129
miR-137	34

Table 11-4 highlights that out of 193 patients in the study, 129 patients were part of the group, and 34 were part of the miR-137 group. The control group accounts for 79.14% of the total sample of patients, and the miR-137 group accounts for 20.86% of the sample.

Listing 11-4 counts and tabulates event outcomes (see Table 11-4).

Listing 11-4. Event Outcomes

```
class_series_event = df.groupby("E").size()
class_series_event = pd.DataFrame(class_series_event, columns=["Count"])
class_series_event
```

Table 11-4. *Count of Event Outcomes*

	Count
E	
0	7
1	156

Table 11-4 shows that out of 163 patients, only 7 patients failed to survive an event beyond a specific time, and 95.71% of them survived beyond the specified time.

Listing 11-5 counts and tabulates event outcomes per group (see Table 11-5).

Listing 11-5. Event Outcomes per Group

```
class_series_event_per_group = df.groupby(["group","E"]).size()
class_series_event_per_group = pd.DataFrame(class_series_event_per_group,
columns = ["Count"])
class_series_event_per_group
```

Table 11-5. *Count of Event Outcome per Group*

		Count
group	E	
control	0	7
	1	122
miR-137	1	34

Table 11-5 shows that all 34 of the patients who were part of the miR-137 group survived an event beyond the specific time. In the control group, 122 of the patients survived and 7 failed to survive.

Data Preprocessing

Listing 11-6 splits the data into the miR-137 group and the control group.

Listing 11-6. Split Data

```
miR_137 = df.loc[df.group == "miR-137"]
control_group = df.loc[df.group == "control"]
```

After splitting the data into two, we repurpose the data into the required format. See Listing 11-7.

Listing 11-7. Repurpose Data

```
T1 = miR_137["T"]
E1 = miR_137["E"]
T2 = control_group["T"]
E2 = control_group["E"]
```

Descriptive Statistics

Remember, in survival analysis, there are no strict assumptions of linearity and normality; we do not use descriptive analysis to test assumptions, but to understand the underlying structure of the data. We know that we are dealing with skewed data. Listing 11-8 returns descriptive statistics (see Table 11-6).

Listing 11-8. Descriptive Statistics

```
pd.DataFrame(miR_137["T"].describe()).transpose()
```

Table 11-6. *MiR-137 Descriptive Statistics*

	Count	mean	std	min	25%	50%	75%	max
T	34.0	25.705882	13.358755	6.0	16.0	26.0	29.0	62.0

Table 11-6 shows that the minimum number of years a patient takes part in the study is 6, and the maximum number of years is 62. The mean number of years a patient takes part in the study is 26.

Survival Table

We widely recognize the life table as the survival table. A survival table is a table that provides information about the activities of the patients while participating in a study. Table 11-7 highlights the contents of a survival table.

Table 11-7. Contents of a Survival Table

Content	Description
Removed	The number of subjects who die (or suffer the event of interest) during interval *t*
Observed	The number of subjects who are event-free and thought of as being at risk during interval *t*
Censored	The number of participants who leave the study during interval *t*
Entrance	The number of subjects who entered during interval *t*
At-risk	The aggregate number of subjects at risk during interval *t*

Listing 11-9 applies the survival_from_event() method to create a survival table (see Table 11-8).

Listing 11-9. Survival Table

```
table = survival_table_from_events(T1,E1)
table.head()
```

Table 11-8. *Survival Table*

	removed	observed	censored	entrance	at_risk
event_at					
0.0	0	0	0	34	34
6.0	1	1	0	0	34
9.0	3	3	0	0	33
13.0	3	3	0	0	30
15.0	2	2	0	0	27

Table 11-8 tells us the following:

- In the first year patients entered the study. By default, we consider all patients under risk.
- At year 6, one patient died, and we thought another to be at risk.

- At year 9, three patients died, and we thought another three to be at risk.
- At year 13, we reported three deaths and three patients at risk.
- By year 15, we reported two deaths and two patients at risk. There were 27 patients left in the miR-137 group.

The Kaplan-Meier Model

The Kaplan-Meier (KM) model is a nonparametric method used to estimate the survival probability from the observed survival times. It measures a binary variable using time. We also recognize the model as a product limit model. When we use the model, we look at the number of patients dying in a specified period, considering that they have already survived. To understand how this model works, we explain the survival function underneath. Listing 11-10 completes the Kaplan-Meier models.

Listing 11-10. Finalize the Kaplan-Meier Models

```
kmf_miR_137 = KaplanMeierFitter().fit(T1,E1)
kmf_control_group = KaplanMeierFitter().fit(T2,E2)
```

After developing the models, we find the estimates of the confidence interval.

Confidence Interval

A confidence interval (CI) represents the probability of a value falling within a specific range. Table 11-9 shows the probability of the survival estimates falling in a specific range at a certain period. The first column shows the timeline, the second column shows the lower limit of the CI, and the third column shows the upper limit of the CI. See Listing 11-11.

Listing 11-11. Kaplan-Meier Estimate Confidence Interval

```
kmf_miR_137_ci = kmf_miR_137.confidence_interval_
kmf_control_group_ci = kmf_control_group.confidence_interval_
kmf_miR_137_ci.head()
```

Table 11-9. Kaplan-Meier Estimate Confidence Interval

	KM_estimate_lower_0.95	KM_estimate_upper_0.95
0.0	1.000000	1.000000
6.0	0.809010	0.995804
9.0	0.716269	0.954139
13.0	0.616102	0.896087
15.0	0.552952	0.852504

Table 11-9 shows, if we were to reproduce the study 95% of the time.

- At year 0, the survival probability of patients in the miR-137 group is 100%.
- At year 6, the survival probability of patients in the miR-137 group is between 80.90% and 99.58%.
- At year 9, the survival probability of patients in the miR-137 group is between 71.27% and 95.41%.
- At year 13, the survival probability of patients in the miR-137 group is between 61.61% and 89.61%.
- At year 13, the survival probability of patients in the miR-137 group is between 55.95% and 85.25%.

Cumulative Density

Cumulative density represents the probability of failure occurring at a specific time *t*. It estimates the cumulative density. We use the formula in Equation 11-1 to estimate the unconditional failure rate.

$$\hat{f} = \frac{d_i}{\left\{ \frac{d_i}{t_{i+1} - t_i} \right\}} \qquad \text{(Equation 11-1)}$$

Here, d_i represents the number of events, and t is the survival time. See Listing 11-12 and Table 11-10.

Listing 11-12. Kaplan-Meier Cumulative Density Estimates

```
kmf_c_density = kmf_miR_137.cumulative_density_
kmf_control_group_c_density = kmf_control_group.cumulative_density_
kmf_c_density.head()
```

Table 11-10. *Kaplan-Meier Cumulative Density Estimates*

	KM_estimate
timeline	
0.0	0.000000
6.0	0.029412
9.0	0.117647
13.0	0.205882
15.0	0.264706

To make sense of the cumulative density estimates, we plot the Kaplan-Meier cumulative density curve.

Cumulative Density Curve

We use a cumulative density curve to understand the median value and the interquartile range. When we plot the curve, the cumulative frequency is on the x-axis, and the upper-class boundary of each interval is on the y-axis. Listing 11-13 and Figure 11-1 show the Kaplan-Meier cumulative density curve. From the curve, we can identify the median quartile and interquartile.

Listing 11-13. Kaplan-Meier Cumulative Density Curve

```
kmf_miR_137.plot_cumulative_density(color="navy", label="miR-137")
kmf_control_group.plot_cumulative_density(color="green", label="control")
plt.xlabel("Time")
plt.ylabel("Cumulative Incidence")
plt.legend(loc=2)
plt.show()
```

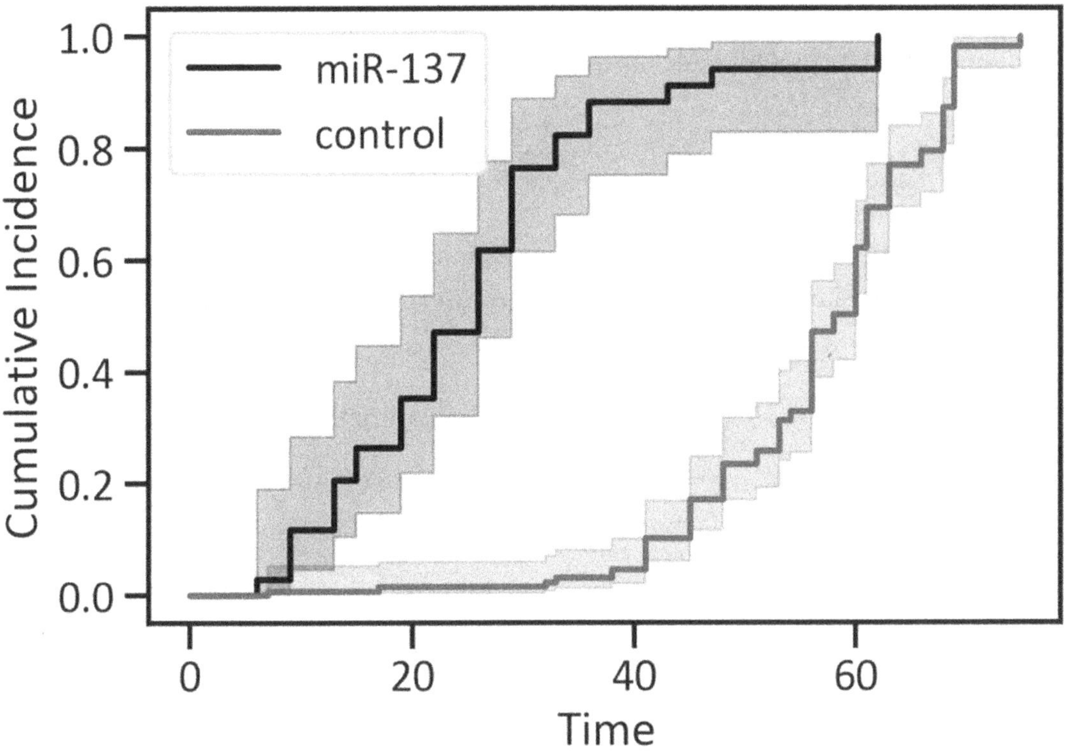

Figure 11-1. *Kaplan-Meier cumulative density curve*

Figure 11-1 shows the following:

- At year 0, the unconditional failure rate is 0%.
- At year 6, the unconditional failure rate is 2.94%.
- At year 9, the unconditional failure rate is 11.17%.
- At year 13, the unconditional failure rate is 20.59%.
- At year 15, the unconditional failure rate is 26.47%.

Survival Function

In survival analysis, event outcomes and follow-up times are used to estimate the survival function. The survival function of time *t* is estimated (in the uncensored case) by the equation in Equation 11-2.

$$\hat{S} = \frac{number\ of\ patients\ surviving\ beyond\ time\ t}{n} \qquad \text{(Equation 11-2)}$$

Here, n is the number of patients in the study (that is, at time zero). We express the formula as shown in Equation 11-3.

$$\hat{S}(t_i) = \frac{(n - (d_0 + d_1 + \ldots + d_i))}{n} \quad \text{(Equation 11-3)}$$

Here, d_i is the number of events, and n is the number of individuals who survived.

Listing 11-14 estimates and tabulates Kaplan-Meier function estimates (see Table 11-11).

Listing 11-14. Kaplan-Meier Survival Function Estimates

```
kmf_survival_function = kmf_miR_137.survival_function_
kmf_control_group_survival_function = kmf_control_group.survival_function_
kmf_survival_function.head()
```

Table 11-11. *Kaplan-Meier Survival Function Estimates*

	KM_estimate
timeline	
0.0	1.000000
6.0	0.970588
9.0	0.882353
13.0	0.794118
15.0	0.735294

Survival Curve

A survival curve is the plot of $\hat{S}(t_i)$ versus t. In a survival curve, the x-axis represents time in years, and the y-axis represents the probability of surviving or the proportion of people surviving. Listing 11-15 plots the survival curve (see Figure 11-2).

Listing 11-15. Kaplan-Meier Survival Curve

```
kmf_miR_137.plot_survival_function(color="navy", label="miR-137")
kmf_control_group.plot_survival_function(color="green",label="control")
plt.xlabel("Time")
plt.ylabel("Survival Probability")
plt.legend(loc=3)
plt.show()
```

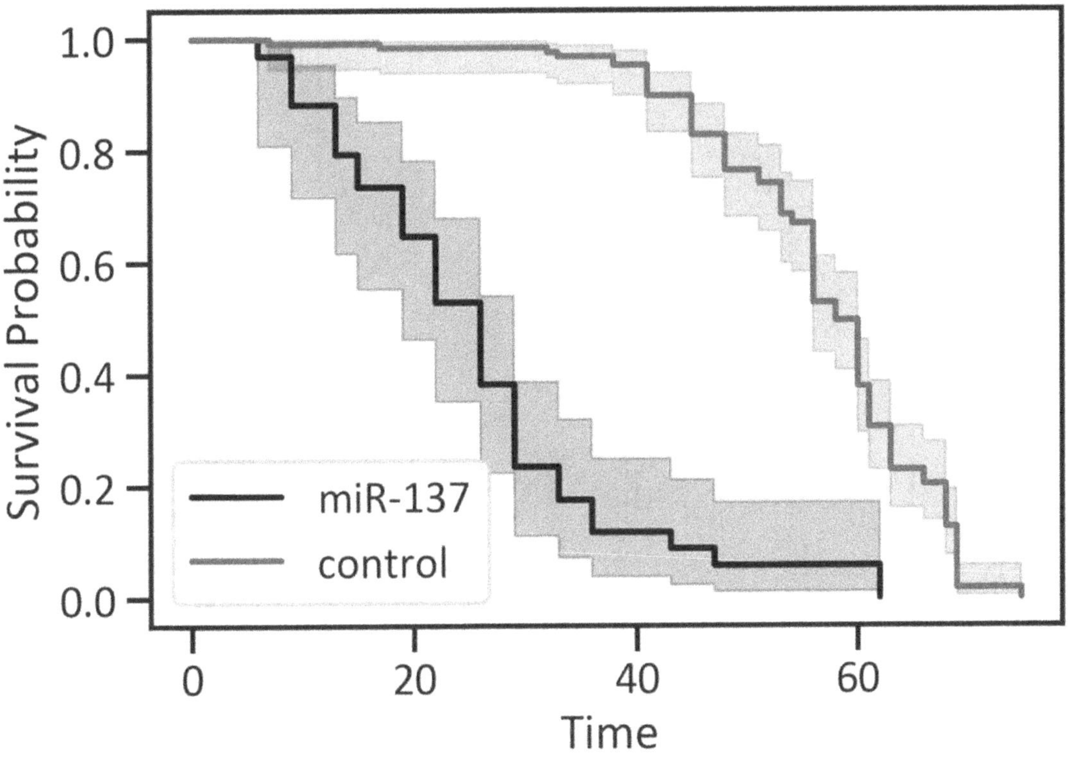

Figure 11-2. *Kaplan-Meier survival curve*

Figure 11-2 shows that there is a major difference between the survival curve of the miR-137 group and the control group. The miR-137 group's survival curve sharply declines at year 9. Meanwhile, the control group's survival curve only starts declining at year 45. Listing 11-16 plots the survival curve with confidence intervals (see Figure 11-3).

Listing 11-16. Kaplan-Meier Curve with Confidence Interval

```
kmf_miR_137.plot_survival_function(color="navy", label="miR-137")
kmf_control_group.plot_survival_function(color="green", label="Control")
plt.plot(kmf_miR_137.confidence_interval_, color="red", label="95% CI")
plt.plot(kmf_control_group.confidence_interval_, color="red")
plt.xlabel("Time")
plt.ylabel("Survival Probability")
plt.legend(loc=3)
plt.show()
```

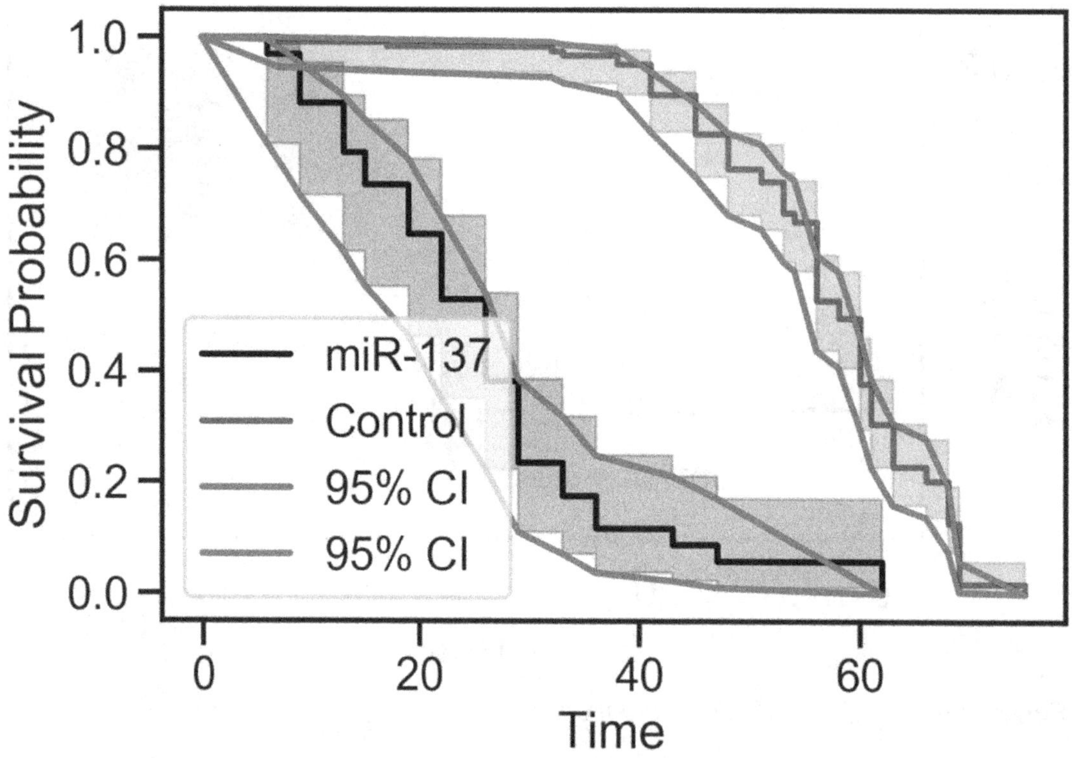

Figure 11-3. *Kaplan-Meier survival curve with confidence intervals*

Figure 11-3 tells us the following:

- At year 0, the probability of a patient surviving is 100%.
- At year 6, the probability of a patient surviving is 97.06%.
- At year 9, the probability of the patient surviving is 88.24%.

- At year 13, the probability of the patient surviving is 79.41%.
- At year 15, the probability of the patient surviving is 73.53%.

Listing 11-17 finds the median survival time.

Listing 11-17. Median Survival Time

```
kmf_miR_137.median_survival_time_
26.0
```

The median survival time is 26 years. Listing 11-18 applies the `predict()` method to predict future instances of event outcomes.

Listing 11-18. Predicted Event Outcome

```
kmf_miR_137.predict(30)
0.23529411764705874
```

When the survival time is 30 years, the event outcome is 0. After seeing how the model predicts instances, we look at how the model performs.

Evaluate the Kaplan-Meier Model

After estimating the survival probabilities of the two groups, we must find whether the death generation of both groups is equal. Unlike other nonparametric models, the Kaplan-Meier model does not have sophisticated model evaluation metrics. The best way to test the model involves finding out whether groups share a similar survival curve. We express the hypothesis as follows:

- *Null hypothesis:* All groups share the same survival curve.
- *Alternative hypothesis:* All groups do not share the same survival curve.

Listing 1-19 returns the log-rank results (see Table 11-12). Listing 11-20 shows the p-value.

Listing 1-19. Log-Rank Test Results

```
from lifelines.statistics import logrank_test
results = logrank_test(T1, T2, event_observed_A=E1, event_observed_B=E2)
results.print_summary()
```

Table 11-12. Log-Rank Test Results

t_0		-1
null_distribution		chi squared
degrees_of_freedom		1
test_name		logrank_test
	test_statistic	p
0	122.25	<0.005

Listing 11-20. P-Value

```
print(results.p_value)
2.0359832222855426e-28
```

The previous findings show that the p-value is greater than 0.05. We reject the null hypothesis in favor of the alternative hypothesis. All groups do not share the same survival curve.

Conclusion

This chapter covered the most popular nonparametric survival analysis model known as the Kaplan-Meier model. We estimated the survival probabilities of a patient surviving an event beyond a specific time. Before developing the model, we split the data into two groups, and we then created a survival table for the group under investigation. We plotted the cumulative density curve and survival curve to make sense of the estimates. Thereafter, we tested whether the two groups share the same survival curve using the log-rank test. After studying the test statistic, we found the model is reliable, and we can use it to predict future instances of event outcomes.

CHAPTER 12

Neural Networks

This chapter introduces a subfield of machine learning frequently recognized as deep learning. First, it introduces different artificial neural networks. Second, we cover back propagation and forward propagation. Third, it presents different activation functions. Last, it builds and test a Restricted Boltzmann Machine and a multilayer perceptron using the SciKit-Learn package, followed by deep belief networks using the Keras package. To install Keras on the Python environment, use `pip install Keras` and on the conda environment use `conda install -c conda-forge keras`.

An artificial neural network is an interconnected group of nodes that retrieves and processes input values using different weights and biases across layers until there is an output value. A neural network comprises several layers with configured nodes. Each node in a hidden layer and output layer contains its own classifier. Nodes in the visible layer retrieve input values, thereafter, the activation happens. Table 12-1 highlights several types of neural networks and their applications.

Table 12-1. *Types of Neural Networks and their Applications*

Neural Network	Application
Restricted Boltzmann Machine or Autoencoder	Unlabelled data, variable extraction, and pattern recognition
Recursive Neural Tensor Network or Recurrent Neural Network	Text processing[1]
Deep Belief Net or Convolutional Neural Network	Image recognition
Recurrent Neural Network	Speech recognition
Multilayer Perceptron orDeep Belief Network	Classification and Time Series Analysis

[1] `https://www.tensorflow.org/guide/keras/rnn`

© Tshepo Chris Nokeri 2021
T. C. Nokeri, *Data Science Revealed*, https://doi.org/10.1007/978-1-4842-6870-4_12

Forward Propagation

Forward propagation involves nodes in the visible layer receiving and processing input values, and transmitting them to nodes in the subsequent layers until an output value is produced. Although each layer retrieves the same input value, they do not transmit the same value.

Back Propagation

Back propagation is the reverse of forward propagation. In back propagation, the network estimates the gradient in reverse, and transforms weights of error rates of the preceding epoch. This process is not memory efficient, and it also results in poor model performance.

Cost Function

Chapter 1 introduced residual analysis, thus understanding the cost function should be relatively easy. A cost function estimates the differences between actual values and predicted values. It adjusts weights and biases until it finds the lowest value. We use cost function and loss function interchangeably.

Gradient

A gradient represents the rate at which the cost changes weights and bias. We calculate the gradient during training. It is used to. A neural network multiples gradients of preceding layers to find the gradient of subsequent layers. To optimize the gradient, we use gradient descent. There three types of gradient descent method, namely 1) batch gradient descent - estimates gradient of cost function to the parameters of the whole training data, 2) stochastic gradient descent - updates parameter for each independent variable and dependent variable, 3) mini-batch gradient descent - applies batch gradient descent and stochastic gradient descent to perform a mini-batch of the training data.

Vanishing Gradient

If the gradient is large, we train the network fast, and if the gradient is small, the training becomes slow. At most, the first layer has a small gradient, and subsequent layers have larger gradients. A traditional machine learning model learns at a slow pace and makes a lot of errors when predicting classes. Artificial neural networks fill this gap by scaling the gradient.

Activation Function

An activation function adds non-linearity to the network. An activation function enables back propagation. Excluding an activation function reduces the neural network to a linear regression model. Underneath, we discuss these activation functions. There are other functions[2] we may use.

The Sigmoid Activation Function

Chapter 5 introduced the logistic function and how it estimates values. We also recognize the logistic function as the sigmoid activation. It triggers output values between 0 and 1. We primarily use the sigmoid activation function to solve binary classification problems. We express the sigmoid function as:

$$Sigmoid = \frac{1}{1+e^{-z}} \quad \text{(Equation 12-1)}$$

The activation function feeds forward neural networks that require positive output values only.

[2] https://keras.io/api/layers/activations/

The Tangent Hyperbolic Activation Function

The tangent hyperbolic (tanh) activation function extends the sigmoid activation function. It triggers output values between -1 and 1 and enables the network to center the mean value to 0. We express the function as:

$$Tanh = \frac{e^{t} - e^{-t}}{e^{t} - e^{-t}} \qquad \text{(Equation 12-2)}$$

The tanh activation fits a tangent hyperbolic curve to the data. Figure 12-1 depicts the standard tanh function.

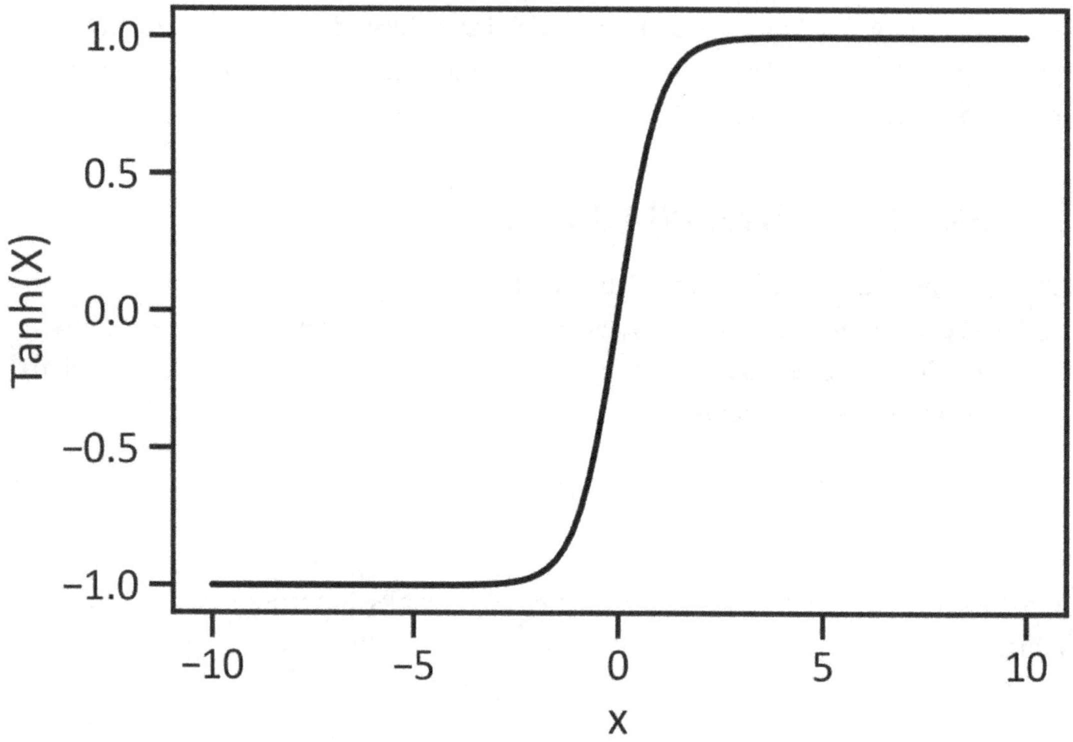

Figure 12-1. *Tanh Function*

Notice the output value is between -1 and 1 and the mean value is centered to 0.

Rectified Linear Units Activation Function

ReLU is the default activation function in CNN and multilayer perceptron. ReLU helps models to learn more fast and its performance is better. Unlike the sigmoid and tanh activation functions, the ReLu activation does not saturate the data to -1, 0, or 1. Rather, it moves onward until it retrieves an optimal value. It helps solve the vanishing gradient problem. We express the function as:

$$f(x) = \max(0, x) \quad \text{(Equation 12-3)}$$

Figure 12-2 depict the standard ReLu function.

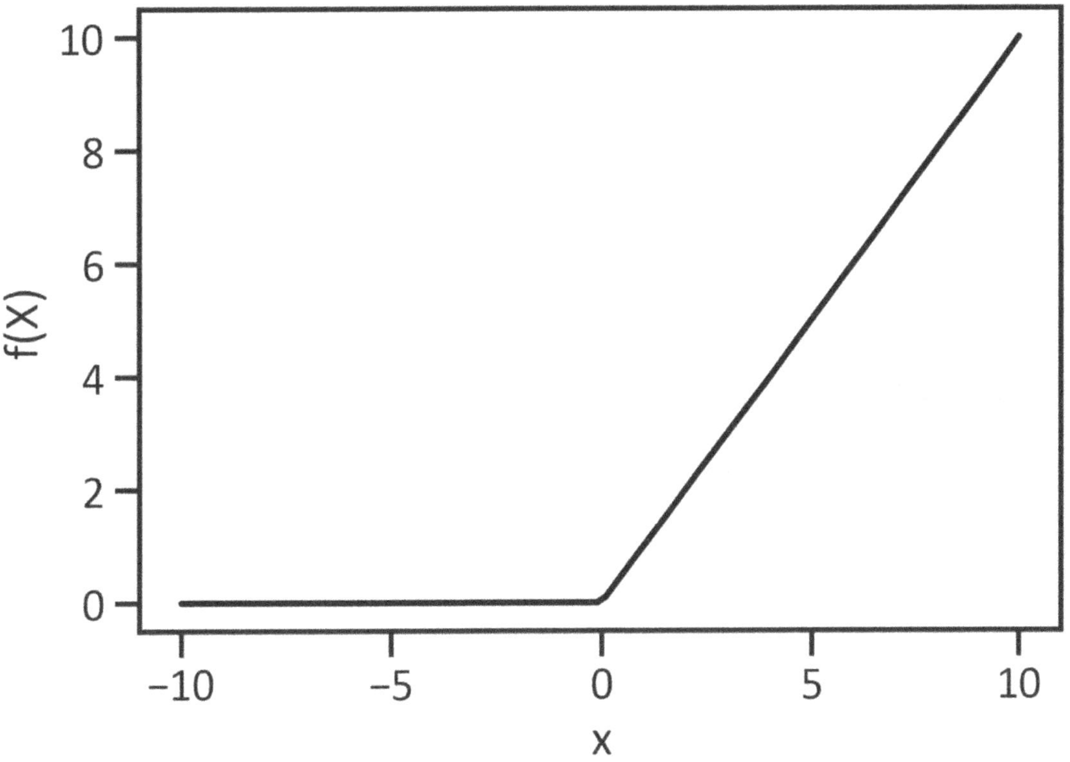

Figure 12-2. *ReLu Function*

Notice the output value is not constrained to a specific range.

Loss Function

Loss is a standard metric for examining the underlying performance of a model. A neural network applies a loss function to determine whether it should improve its learning process per epoch. An increasing loss shows that a model is a skillful learner. The most common loss functions for continuous dependent variables include Mean Squared Error (MSE), Mean Absolute Error (MAE), Mean Absolute Percentage Error (MAPE), and Mean Squared Logarithmic Error (MSLE). Loss functions for categorical outcomes include binary cross-entropy for binary classification and categorical cross-entropy for multiclass classification.

Optimization

There are many methods for model performance optimization, like SGD, RMSProp, ADAGRAD, Adadelta, and Adam. This chapter applies the Adam optimization method[3] because it works better in minimizing the cost function during training. Adam stands for Adaptive Movement Estimation. It considers preceding gradients in momentum and lowers the learning rate.

Bernoulli Restricted Boltzmann Machine

We use the Restricted Boltzmann Machine (RBM) to address the vanishing gradient problem. An RBM is a shallow neural network that comprises the visible layer and the hidden layer. It connects each node in the visible layer to every node in the hidden layer. In addition, it applies weight and biases to understand the underlying relationship between variables.

[3] `https://keras.io/api/optimizers/adam/`

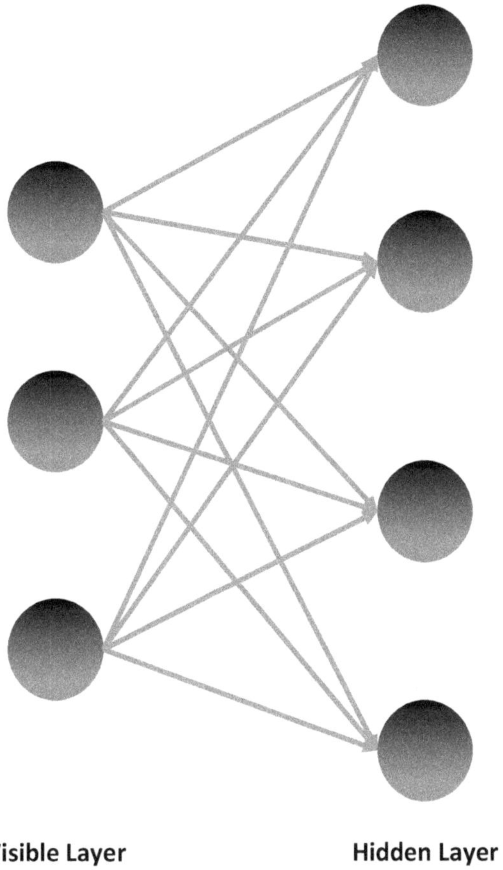

Figure 12-3. *Restricted Boltzmann Machine*

Figure 12-3 shows an RBM with 3 nodes at the visible layer and 4 nodes at the hidden layer.

Train a Logistic Classifier

Listing 12-1 trains the logistic classifier.

Listing 12-1. Finalize the Logistic Classifier

```
from sklearn.linear_model import LogisticRegression
logreg = LogisticRegression(dual= False,
                            fit_intercept= True,
                            max_iter= 10,
                            n_jobs= -5,
```

```
                            penalty= 'l2',
                            tol=0.0001,
                            warm_start= False)
logreg.fit(x_train, y_train)
```

Pipeline

Listing 12-2 creates a pipeline for the logistic classifier and RBM classifier.

Listing 12-2. Complete the Pipeline

```
from sklearn.neural_network import BernoulliRBM
from sklearn.pipeline import Pipeline
rbm = BernoulliRBM()
classifier = Pipeline(steps=[("rbm", rbm), ("logreg", logreg)])
classifier.fit(x_train, y_train)
```

Table 12-2 highlights key classification evaluation metrics such as accuracy, precision, recall, and others.

Listing 12-3. Classification Report

```
y_pred = classifier.predict(x_test)
creportbm = pd.DataFrame(metrics.classification_report(y_test, y_pred,
output_dict=True)).transpose()
creportbm
```

Table 12-2. *Classification Report*

	precision	**recall**	**f1-score**	**support**
0	0.829787	0.728972	0.776119	107.000000
1	0.516667	0.659574	0.579439	47.000000
accuracy	0.707792	0.707792	0.707792	0.707792
macro avg	0.673227	0.694273	0.677779	154.000000
weighted avg	0.734224	0.707792	0.716094	154.000000

The Bernoulli RBM classifier is accurate 70.78% of the time when predicting classes. The accuracy score is lower than that of the classifiers we covered in the preceding chapters. The classifier is precise 82.98% of the time when predicting class 0 and precise 51.67% of the time when predicting class 1.

Multilayer Perceptron using SciKit-Learn

At most, the single perceptron model struggles to capture the underlying structure of the data, resulting in poor model performance. The Multilayer Perceptron (MLP) model to address this. It trains, weights, and thresholds a set of random values, then estimates and controls change using the activation function. It encompasses multiple layers; there must be at least three layers, namely the visible layer, the hidden layer, and the output layer.

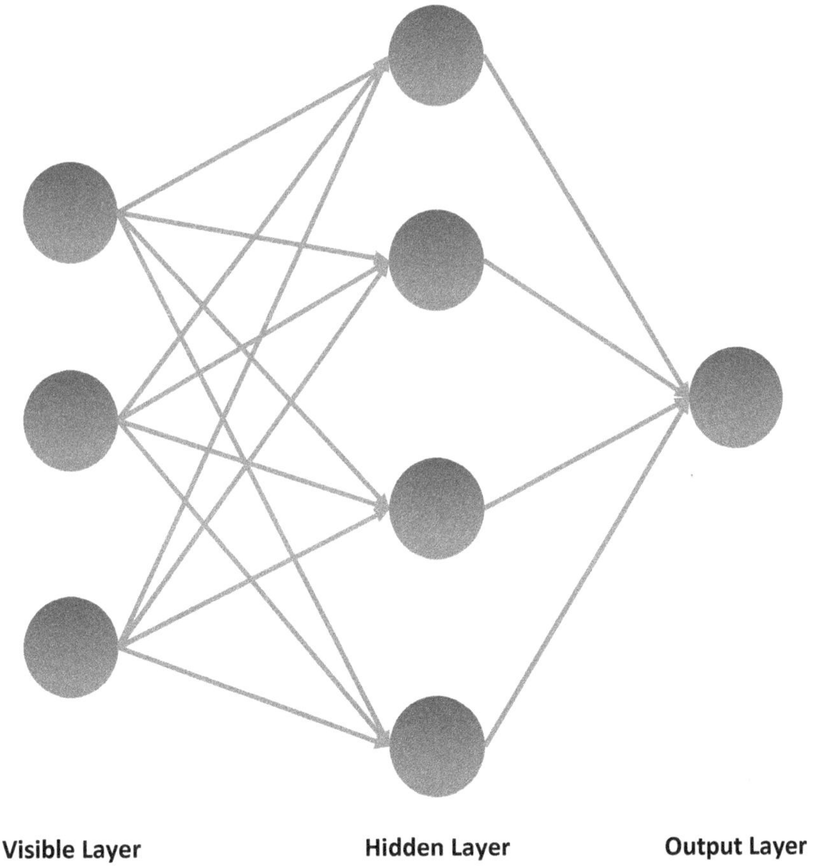

Figure 12-4. *Multilayer Perceptron*

Figure 12-4 shows the MLP with 3 nodes at the visible layer, 4 nodes at the hidden layer, and an output layer with only one possible outcome.

Listing 12-4. Finalize the MLP Classifier

```
from sklearn.neural_network import MLPClassifier
mlp = MLPClassifier()
mlp.fit(x_train, y_train)
```

Table 12-3 provides the model's accuracy score, precision score, recall, and other key evaluation metrics

Listing 12-5. Classification Report

```
y_predmlp = mlp.predict(x_test)
creportmlp = pd.DataFrame(metrics.classification_report(y_test, y_predmlp,
output_dict=True)).transpose()
creportmlp
```

Table 12-3. *Classification Report*

	precision	recall	f1-score	support
0	0.848214	0.887850	0.867580	107.000000
1	0.714286	0.638298	0.674157	47.000000
accuracy	0.811688	0.811688	0.811688	0.811688
macro avg	0.781250	0.763074	0.770869	154.000000
weighted avg	0.807340	0.811688	0.808548	154.000000

The MLP classifier is more accurate and precise compared to the RBM classifier.

Compare the Bernoulli RBM and MLP ROC Curves

Figure 12-5 shows how skillful the classifiers are in distinguishing between classes.

Listing 12-6. ROC Curves

```
y_pred_probarbm = classifier.predict_proba(x_test)[::, 1]
y_pred_probamlp = mlp.predict_proba(x_test)[::, 1]
aucrbm = metrics.roc_auc_score(y_test, y_pred_probarbm)
aucmlp = metrics.roc_auc_score(y_test, y_pred_probamlp)
fprrbm, tprrbm, _ = metrics.roc_curve(y_test, y_pred_probarbm)
fprmlp, tprmlp, _ = metrics.roc_curve(y_test, y_pred_probamlp)
plt.plot(fprrbm, tprrbm, label="auc: "+str(aucrbm), color="gray")
plt.plot(fprmlp, tprmlp, label="auc: "+str(aucmlp), color="black")
plt.plot([0, 1], [0, 1], color="red")
plt.xlabel("Specificity")
plt.ylabel("Sensitivity")
plt.legend(loc=4)
plt.show()
```

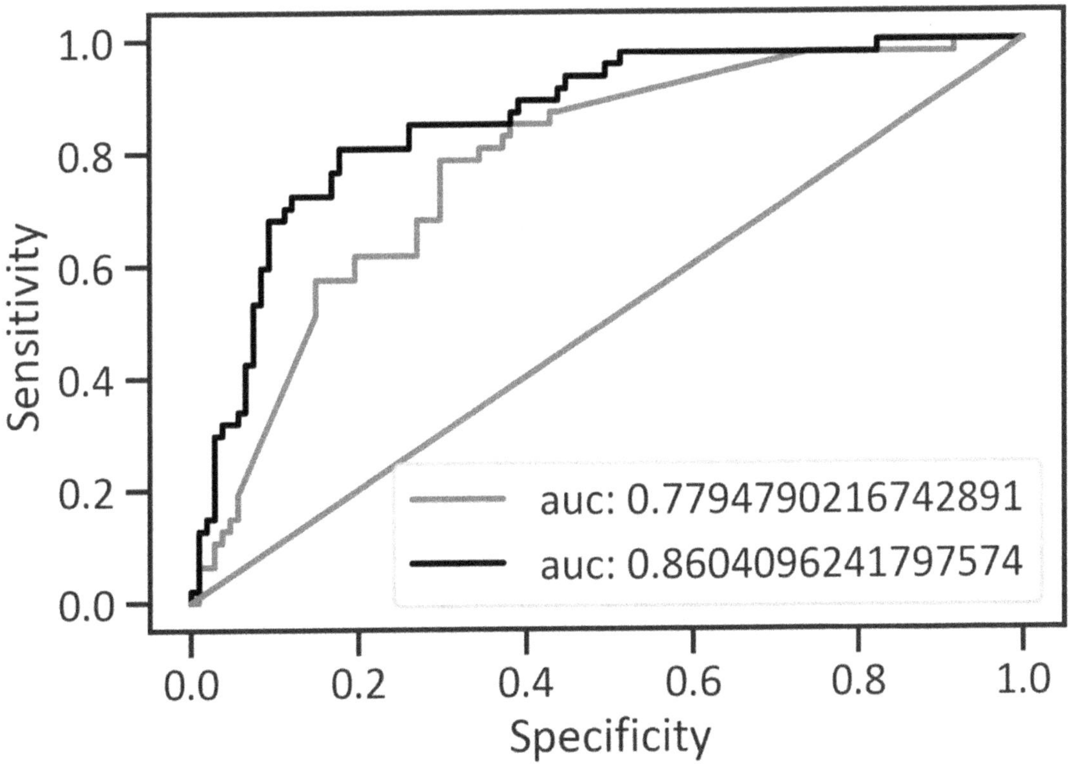

Figure 12-5. *ROC Curves*

In Figure 12-5, the curve of the MLP classifier are highlighted in black and that of the Bernoulli RBM classifier is highlighted in gray. The Bernoulli RBM's curve is closer to the 45-degree line when compared to that of the MLP classifier. None of the curves are close to the right-hand-side of the border. They do not show the characteristics of a perfect ROC curve. However, the MLP classifier has an AUC score of 0.86 (greater than most classifiers covered in the chapters). Meanwhile, the AUC score of the Bernoulli RBM classifier is less than 0.80 (less than the score of all classifiers covered in the preceding chapters).

Compare Bernoulli RBM and MLP Precision-Recall Curves

Figure 12-6 how the classifiers trade-off precision and recall across different thresholds.

Listing 12-7. Precision-Recall Curves

```
precisionrbm, recallrbm, thresholdrbm = metrics.precision_recall_curve(
y_test, y_predrbm)
precisionmlp, recallmlp, thresholdmlp = metrics.precision_recall_curve(
y_test, y_predmlp)
apsrbm = metrics.average_precision_score(y_test, y_predrbm)
apsmlp = metrics.roc_auc_score(y_test, y_predmlp)
plt.plot(precisionrbm, recallrbm, label="aps: "+str(apsrbm), color="gray")
plt.plot(precisionmlp, recallmlp, label="aps: "+str(apsmlp), color="black")
plt.axhline(y=0.5, color="red")
plt.xlabel("Recall")
plt.ylabel("Precision")
plt.legend(loc=3)
plt.show()
```

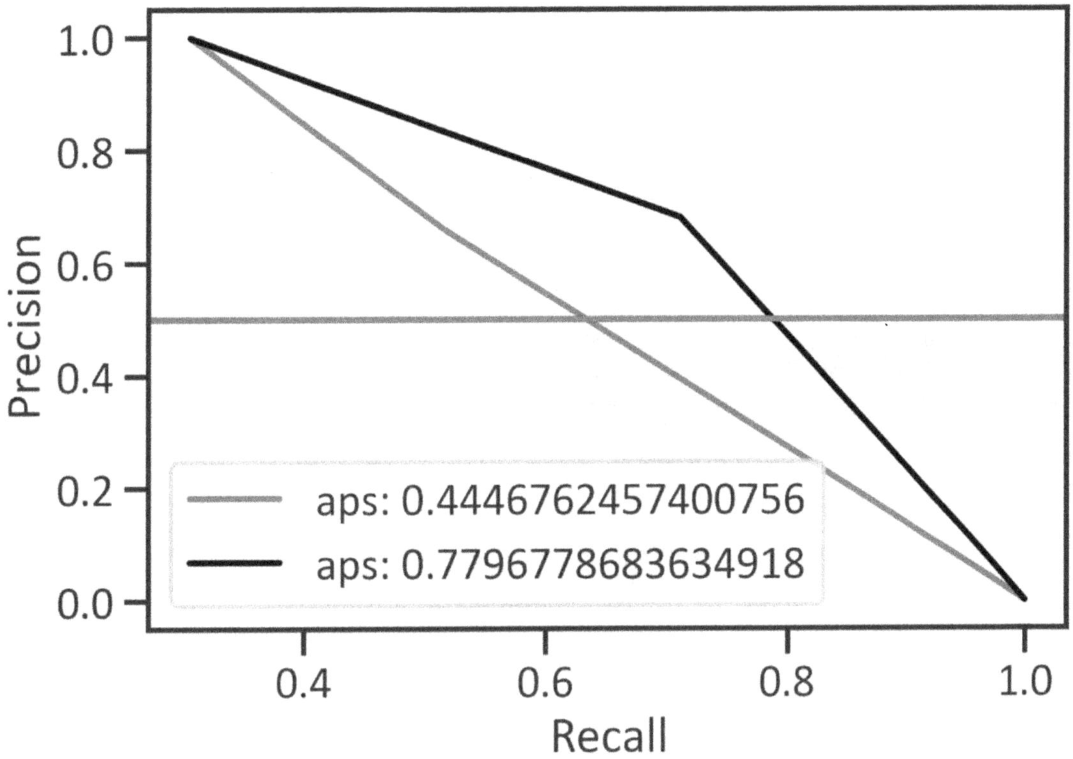

***Figure 12-6.** Precision-Recall Curves*

All curves do not follow the top border, nor approach the top-right border. Rather, they quickly reach the 1.0 point. The MLP classifier is more precise than the Bernoulli RBM classifier on average.

Deep Belief Networks using Keras

A deep belief network is a combination of multiple RBMs. The network is an alternative to backward propagation. The structure is like that of the MLP network. However, they differ in training; a hidden layer of one RBM is a visible layer of another RBM.

Split Data into Training, Test Data and Validation Data

Previously, we split the data into training data and test data, whereby 80% of the data is for training, and 20% of the data is for testing. From now on, we allocate a certain portion of the training data for validation. First, we split the data into training data and test data using the 80/20 ratio. Last, we further split the training data to allocate a portion of 10% for validation. Figure 12-7 shows how we split the data into three sets.

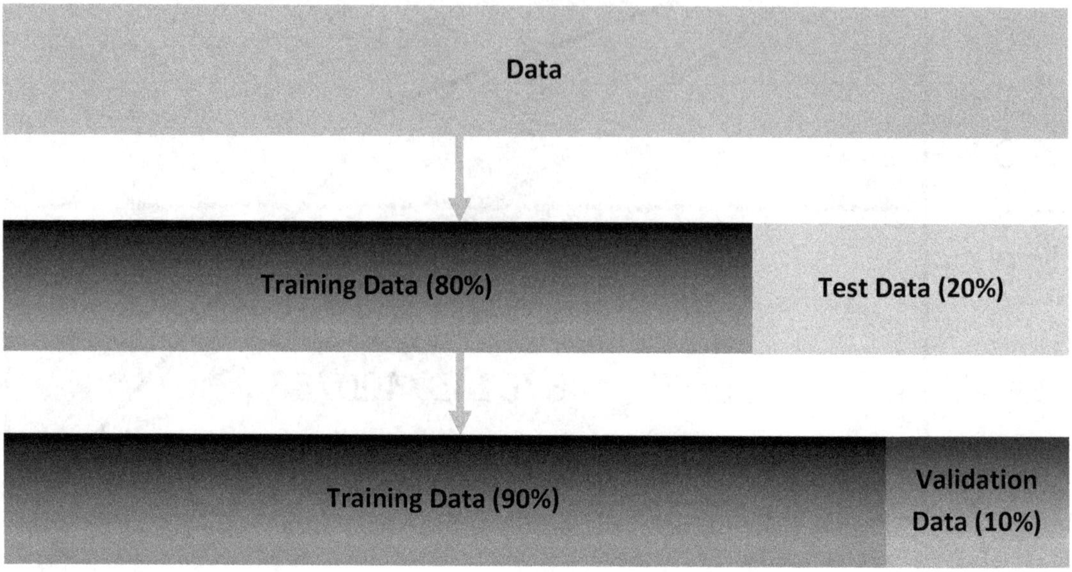

Figure 12-7. *Data Splits*

We use the `training_test_split()` method to split the data into training data, test data, and validation data.

Listing 12-8. Split the Data into Training Data, Test Data and Validation Data

```
x_train, x_test, y_train, y_test = train_test_split(x, y, test_size=0.2,
random_state=0)
x_train, x_val, y_train, y_val = train_test_split(x_train, y_train,
test_size=0.2, random_state=0)
x_train = scaler.fit_transform(x_train)
x_test = scaler.transform(x_test)
```

Afterward, we import the Keras package.

Listing 12-9. Import the Keras Package

```
import tensorflow as tf
from keras import Sequential, regularizers
from keras.layers import Dense, Dropout
from keras.wrappers.scikit_learn import KerasClassifier
```

The first network comprises a visible layer with the sigmoid activation function. There are also two hidden layers with 8 neurons and the ReLu activation function.

Listing 12-10. Build Model 1's Architecture

```
def create_dnn_model1(optimizer="adam"):
    model1 = Sequential()
    model1.add(Dense(8, input_dim=8, activation="sigmoid"))
    model1.add(Dense(8, activation="relu"))
    model1.add(Dense(1, activation="relu"))
    model1.compile(loss="binary_crossentropy", optimizer=optimizer,
    metrics=["accuracy"])
    return model1
```

We wrap the architecture of the network using the `KerasClassifier()` method.

Listing 12-11. Wrap Model 1

```
model1 = KerasClassifier(build_fn=create_dnn_model1)
```

Hyperparameter Optimization

Listing 12-12 finds the optimal number of samples to include in training and the number of complete forward and backward passes.

Listing 12-12. Hyperparameter Optimization

```
batch_size = [15, 30, 60]
epochs = [16, 32, 64]
param_grid = {"batch_size":batch_size, "epochs": epochs}
grid_model = GridSearchCV(estimator=model1, param_grid=param_grid)
grid_model.fit(x_train, y_train, validation_data=(x_val, y_val))
print("Best scores: ", grid_model.best_score_, "Best parameters: ",
grid_model.best_params_)
```

Best scores: 0.7535971999168396 Best parameters: {'batch_size': 15, 'epochs': 64}

We use the results above to complete the model.

Finalize Model

We use the `fit()` method to train the model.

Listing 12-13. Finalize Model 1

```
history1 = model1.fit(x_train, y_train, validation_data=(x_val, y_val),
batch_size=15, epochs=64)
history1
```

Table 12-4 provides the model's accuracy score, precision score, recall, and other key evaluation metrics

Listing 12-14. Classification Report

```
y_predmodel1 = model1.predict(x_test)
creportmodel1 = pd.DataFrame(metrics.classification_report(y_test,
y_predmodel1, output_dict=True)).transpose()
creportmodel1
```

Table 12-4. *Classification Report*

	precision	recall	f1-score	support
0	0.862069	0.934579	0.896861	107.000000
1	0.815789	0.659574	0.729412	47.000000
accuracy	0.850649	0.850649	0.850649	0.850649
macro avg	0.838929	0.797077	0.813136	154.000000
weighted avg	0.847945	0.850649	0.845756	154.000000

The first deep belief network shows superior performance, it is more accurate than any other classifier covered.

Regularization

In the second chapter, we introduced the concept of bias-variance trade-off. You can refer back to the chapter. We regularize the deep belief network by using the dropout method or adding a penalty term.

Dropout

We use the dropout method to remove neurons at every layer during training to reduce overfitting. The network estimates the probability that it will drop a neuron in a layer. A dropout rate represents the fraction of the variables dropped in a layer. Listing 12-15 the second deep belief network and drop neurons in layers at a 0.2 rate.

Listing 12-15. Build Model 2's Architecture

```
def create_dnn_model2(optimizer="adam"):
    model2 = Sequential()
    model2.add(Dense(8, input_dim=8, activation="sigmoid"))
    model2.add(Dense(8, activation="relu"))
    model2.add(Dropout(0.2))
    model2.add(Dense(1, activation="relu"))
    model2.compile(loss="binary_crossentropy", optimizer=optimizer,
    metrics=["accuracy"])
    return model2
```

Listing 12-16 wraps the classifier.

Listing 12-16. Wrap Model 2

```
model2 = KerasClassifier(build_fn=create_dnn_model2)
```

Thereafter, we complete the classifier.

Listing 12-17. Finalize Model 2

```
history2 = model2.fit(x_train, y_train, validation_data=(x_val, y_val),
batch_size=15, epochs=64)
history2
```

Table 12-5 provides the model's accuracy score, precision score, recall, and other key evaluation metrics

Listing 12-18. Classification Report

```
y_predmodel2 = model2.predict(x_test)
creportmodel2 = pd.DataFrame(metrics.classification_report(y_test,
y_predmodel2, output_dict=True)).transpose()
creportmodel2
```

Table 12-5. *Classification Report*

	precision	recall	f1-score	support
0	0.844828	0.915888	0.878924	107.000000
1	0.763158	0.617021	0.682353	47.000000
accuracy	0.824675	0.824675	0.824675	0.824675
macro avg	0.803993	0.766455	0.780638	154.000000
weighted avg	0.819902	0.824675	0.818931	154.000000

The dropout method cannot improve the performance of the classier at the rate we specified. However, the classifier outperforms those covered in the preceding chapters.

L1 Regularization

We use L1 regularization to center the data to its central point. This regularization technique makes weak variables produce coefficients equal to zero. It helps reduce noisy variables, consequently optimizing the learning process. Listing 12-19 the architecture of the third deep belief network. It compromises a visible layer with the sigmoid activation function and an L1 penalty term with an alpha of 0.01. There are also two hidden layers with the ReLu activation function.

Listing 12-19. Build Model 3's Architecture

```
def create_dnn_model3(optimizer="adam"):
    model3 = Sequential()
    model3.add(Dense(8, input_dim=8, activation="sigmoid", kernel_
    regularizer=regularizers.l1(0.001), bias_regularizer=regularizers.
    l1(0.01)))
    model3.add(Dense(8, activation="relu"))
```

```
    model3.add(Dense(1, activation="relu"))
    model3.compile(loss="binary_crossentropy", optimizer=optimizer,
    metrics=["accuracy"])
    return model3
```

Listing 12-20 wraps the classifier.

Listing 12-20. Wrap Model 3

```
model3 = KerasClassifier(build_fn=create_dnn_model3)
```

Thereafter, we complete the classifier.

Listing 12-21. Finalize Model 3

```
history3 = model3.fit(x_train, y_train, validation_data=(x_val, y_val),
batch_size=30, epochs=64)
history3
```

Table 12-6 provides the model's accuracy score, precision score, recall, and other key evaluation metrics.

Listing 12-22. Classification Report

```
y_predmodel3 = model3.predict(x_test)
creportmodel3 = pd.DataFrame(metrics.classification_report(y_test,
y_predmodel3, output_dict=True)).transpose()
creportmodel3
```

Table 12-6. *Classification Report*

	precision	recall	f1-score	support
0	0.859649	0.915888	0.886878	107.000000
1	0.775000	0.659574	0.712644	47.000000
accuracy	0.837662	0.837662	0.837662	0.837662
macro avg	0.817325	0.787731	0.799761	154.000000
weighted avg	0.833815	0.837662	0.833702	154.000000

Adding an L1 penalty to the classifier reduces the imbalance. However, it does not improve the accuracy score. The classifier is more precise than all classifiers when predicting class 1.

L2 Regularization

We base L2 regularization on the premise that after normalizing the data, coefficients are small and that the value of *k* increases, coefficients with multicollinearity alter their behavior. This regularization technique is the most suitable for variables with severe correlation. Listing 12-23 builds the architecture of the fourth deep belief network. It compromises a visible layer with the sigmoid activation function and an L2 penalty term with an alpha of 0.01. There are also two hidden layers with the ReLu activation function.

Listing 12-23. Build Model 4's Architecture

```
def create_dnn_model4(optimizer="adam"):
    model4 = Sequential()
    model4.add(Dense(8, input_dim=8, activation="sigmoid", kernel_
    regularizer=regularizers.l2(0.001), bias_regularizer=regularizers.
    l2(0.001)))
    model4.add(Dense(8, activation="relu"))
    model4.add(Dense(1, activation="relu"))
    model4.compile(loss="binary_crossentropy", optimizer=optimizer,
    metrics=["accuracy"])
    return model4
```

Listing 12-24 wraps the classifier.

Listing 12-24. Wrap Model 4

```
model4 = KerasClassifier(build_fn=create_dnn_model4)
```

Thereafter, we complete the classifier.

Listing 12-25. Finalize Model 4

```
history4 = model4.fit(x_train, y_train, validation_data=(x_val, y_val),
batch_size=30, epochs=64)
history4
```

Table 12-6 provides the model's accuracy score, precision score, recall, and other key evaluation metrics.

Listing 12-26. Classification Report

```
y_predmodel4 = model4.predict(x_test)
creportmodel4 = pd.DataFrame(metrics.classification_report(y_test,
y_predmodel4, output_dict=True)).transpose()
creportmodel4
```

Table 12-7. *Classification Report*

	precision	recall	f1-score	support
0	0.850877	0.906542	0.877828	107.000000
1	0.750000	0.638298	0.689655	47.000000
accuracy	0.824675	0.824675	0.824675	0.824675
macro avg	0.800439	0.772420	0.783742	154.000000
weighted avg	0.820090	0.824675	0.820399	154.000000

Adding an L2 penalty term to the deep belief network does not improve the accuracy score. When we carefully compare classification reports of penalized networks, we notice that adding an L1 penalty term reduces the imbalance in the data at the cost of accuracy.

Compare Deep Belief Networks' ROC Curves

Figure 12-8 succinctly summarizes the trade-off between specificity and sensitivity across different probability thresholds (for all models). The closer the curve is to the left-hand-side of the border, then the top border of the space, the more accurate it is.

Listing 12-27. ROC Curves

```
y_pred_probamodel1 = model1.predict_proba(x_test)[::,1]
y_pred_probamodel2 =  model2.predict_proba(x_test)[::,1]
y_pred_probamodel3 = model3.predict_proba(x_test)[::,1]
y_pred_probamodel4 = model4.predict_proba(x_test)[::,1]
```

```
fprmodel1, tprmodel1, _ = metrics.roc_curve(y_test, y_pred_probamodel1)
fprmodel2, tprmodel2, _ = metrics.roc_curve(y_test, y_pred_probamodel2)
fprmodel3, tprmodel3, _ = metrics.roc_curve(y_test, y_pred_probamodel3)
fprmodel4, tprmodel4, _ = metrics.roc_curve(y_test, y_pred_probamodel4)
plt.plot(fprmodel1, tprmodel1, color="navy", label="Model 1")
plt.plot(fprmodel2, tprmodel2, color="orange", label="Model 2")
plt.plot(fprmodel3, tprmodel3, color="green", label="Model 3")
plt.plot(fprmodel4, tprmodel4, color="brown", label="Model 4")
plt.plot([0, 1], [0, 1], color="red")
plt.xlabel("Specificity")
plt.ylabel("Sensitivity")
plt.legend(loc=4)
plt.show()
```

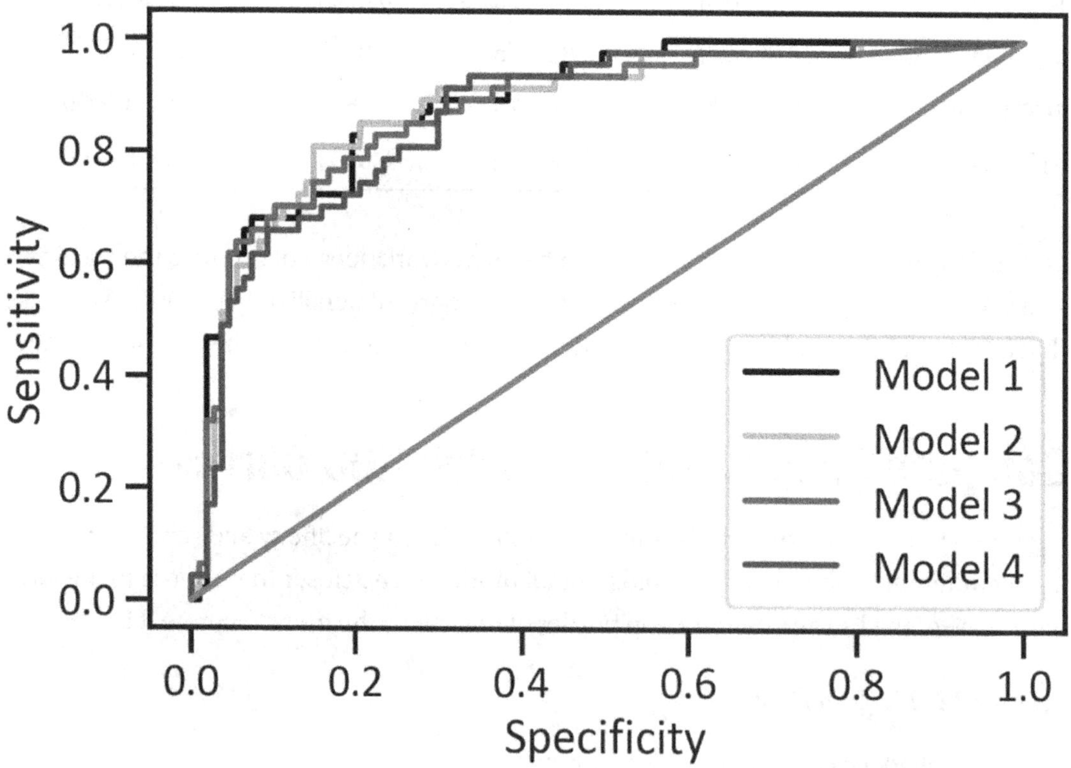

Figure 12-8. *ROC Curves*

Figure 12-8 shows that all the curves show the characteristics of well-behaved curves. They do not follow the left-hand-side of the border nor reach the top of the left-hand-side of the border. Although the curves struggle to do so, they gradually approach the 45-degree line.

Listing 12-28. AUC Scores

```
aucmodel1 = metrics.roc_auc_score(y_test, y_predmodel1)
aucmodel2 = metrics.roc_auc_score(y_test, y_predmodel2)
aucmodel3 = metrics.roc_auc_score(y_test, y_predmodel3)
aucmodel4 = metrics.roc_auc_score(y_test, y_predmodel4)
aucfinal = [[aucmodel1, aucmodel2, aucmodel3, aucmodel3]]
aucfinaldata = pd.DataFrame(aucfinal, columns = ("Model 1",
                                                  "Model 2",
                                                  "Model 3",
                                                  "Model 4"),
                            index=["AUC Score"]).transpose()
aucfinaldata
```

When we construct the ROC curve, we are keenly interested in finding the AUC score. Table 12-8 highlights the AUC scores of all the networks.

Table 12-8. *Deep Belief Networks' AUC Scores*

	AUC Score
Model 1	0.797077
Model 2	0.766455
Model 3	0.787731
Model 4	0.787731

The unregularized network has the greatest AUC score, followed by networks regularized using L1 and L2 penalty terms. The network with dropped neurons is the most mediocre performer.

Compare Deep Belief Networks' Precision-Recall Curves

Our networks are more precise when predicting class 0 than class 1. We must base conclusions about the networks' performance on the trade-off between precision and recall. Listing 12-29 plots a precision-recall curves of all networks (see Figure 12-9).

Listing 12-29. Precision-Recall Curves

```
precisionmodel1, recallmodel1, thresholdmodel1 = metrics.precision_recall_
curve(y_test, y_predmodel1)
precisionmodel2, recallmodel2, thresholdmodel2 = metrics.precision_recall_
curve(y_test, y_predmodel2)
precisionmodel3, recallmodel3, thresholdmodel3 = metrics.precision_recall_
curve(y_test, y_predmodel3)
precisionmodel4, recallmodel4, thresholdmodel4 = metrics.precision_recall_
curve(y_test, y_predmodel4)
plt.plot(precisionmodel1, recallmodel1, label="Model 1", color="navy")
plt.plot(precisionmodel2, recallmodel2, label="Model 2", color="orange")
plt.plot(precisionmodel3, recallmodel3, label="Model 3", color="green")
plt.plot(precisionmodel4, recallmodel4, label="Model 4", color="brown")
plt.axhline(y=0.5, color="red")
plt.xlabel("Recall")
plt.ylabel("Precision")
plt.legend(loc=3)
plt.show()
```

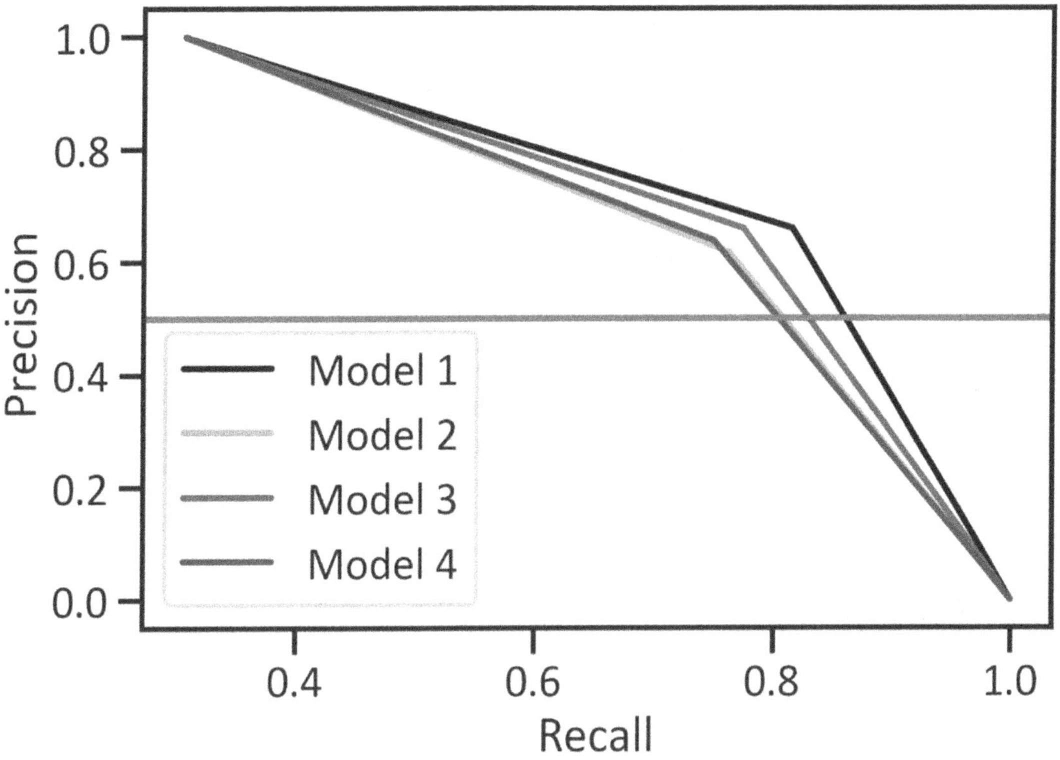

Figure 12-9. *Precision-Recall Curves*

As we progressively increase the recall, the precision of the fourth network drops faster than all models. Figure 12-9 also shows the first network outperforms other networks. Table 12-19 summarizes the precision and recall of all networks using the APS score.

Listing 12-30. APS Score

```
apsmodel1 = metrics.average_precision_score(y_test, y_predmodel1)
apsmodel2 = metrics.average_precision_score(y_test, y_predmodel2)
apsmodel3 = metrics.average_precision_score(y_test, y_predmodel3)
apsmodel4 = metrics.average_precision_score(y_test, y_predmodel4)
apsfinal = [[apsmodel1, apsmodel2, apsmodel3, apsmodel3]]
apsfinaldata = pd.DataFrame(apsfinal, columns = ("Model 1",
                                                  "Model 2",
                                                  "Model 3",
                                                  "Model 4"),
                            index=["APS Score"]).transpose()
apsfinaldata
```

Table 12-9. *APS Score*

	APS Score
Model 1	0.641970
Model 2	0.587768
Model 3	0.615066
Model 4	0.615066

The unregularized network has the greatest APS score, followed by networks regularized using L1 and L2 penalty terms. The network with dropped neurons is less precise.

Training and Validation Loss across Epochs

Figure 12-10 shows how the leading deep belief network learns to compare actual classes and predicted classes.

Listing 12-31. Training and Validation Loss across Epochs

```
plt.plot(history1.history["loss"], color="red", label="Training Loss")
plt.plot(history1.history["val_loss"], color="green", label="Cross-
Validation Loss")
plt.xlabel("Epochs")
plt.ylabel("Loss")
plt.legend(loc=4)
plt.show()
```

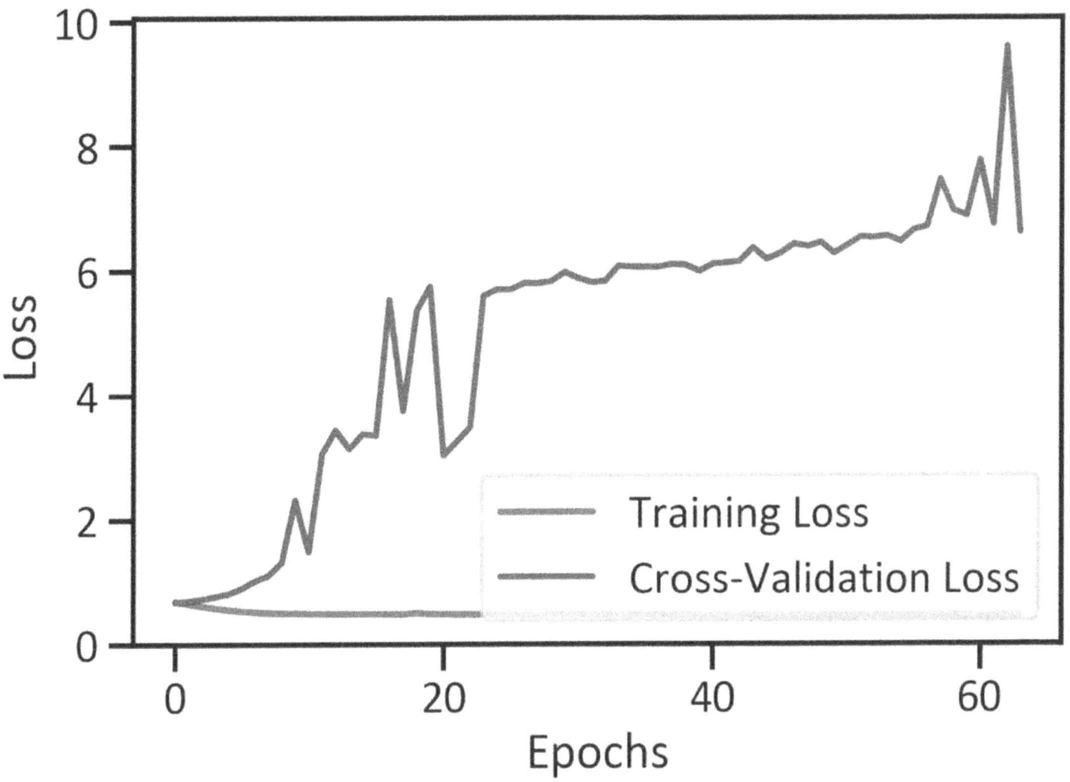

Figure 12-10. *Training and Validation Loss across Epochs*

Figure 12-10 shows that in the 1st epoch, cross-validation loss increases until it reaches its peak at the 60th epoch and begins to drop. The training loss is constantly 0 across epochs.

Training and Validation Accuracy across Epochs

Figure 12-11 shows how the leading deep belief network learns to predict classes correctly.

Listing 12-32. Training and Validation Accuracy across Epochs

```
plt.plot(history1.history["accuracy"], color="red", label="Training
Accuracy")
plt.plot(history1.history["val_accuracy"], color="green", label="Cross-
Validation Accuracy")
```

```
plt.xlabel("Epochs")
plt.ylabel("Accuracy")
plt.legend(loc=4)
plt.show()
```

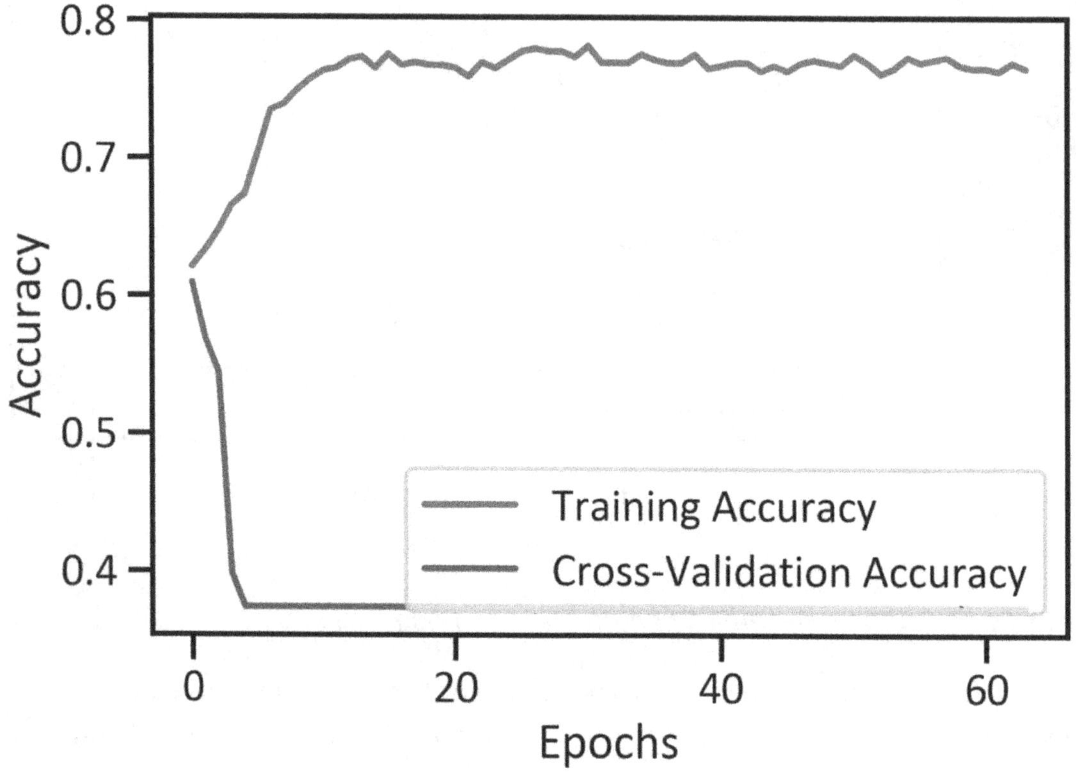

Figure 12-11. *Training and Validation Accuracy across Epochs*

Figure 12-11 highlights that in the 1st epoch, the training accuracy increases to about 80%. Meanwhile, the cross-validation accuracy declines until it reaches 0%, thereafter, it is constantly 0%.

Conclusion

This chapter described artificial neural networks and their applications, and it then explained various activation functions. It solved the same classification problem covered in the preceding chapters using different artificial neural networks such as the Bernoulli RBM classifier, MLP classifier, and deep belief networks.

After properly building their architecture and completing them, we tested their performance using several classification evaluation matrices. We found that the Bernoulli RBM classifier struggles to distinguish between classes. However, the MLP classifier is skillful in distinguishing between classes. All regularized deep belief networks show characteristics of well-behaved classifiers. Moreover, the unregularized deep belief network outperforms the rest.

CHAPTER 13

Machine Learning Using H2O

This chapter concludes a book that familiarizes you with the world of data science. This book covers supervised learning and unsupervised learning, as well as dimension reduction. In addition, it concealed a subfield of machine learning, frequently recognized as deep learning. You might have realized that the field of machine learning is broad. You must be able to engineer data, optimize hyperparameters and develop, test, validate, deploy, and scale models to solve complex problems using machine learning models and deep learning. This typically requires an individual to know and apply different statistical, machine learning, and deep learning models, and some programming techniques.

The demand for data scientists is high, but the supply of data scientists who possess adequate skills is low. Over the past decade, think tanks across the globe have been contributing to well-documented and ground-breaking open-source packages that gallantly help us solve complex problems using machine learning. There is also a wide range of content available in the public domain, like online tutorials, traditional degree courses, online learning courses, published academic research papers, and more. Although the scientific community is attempting to accelerate the adoption AI by make it easily accessible to everyone. Most organizations struggle to unanimously adopt and scale AI solutions.

To get the best of H2O, use H2O Flow. It adequately provides the power of not writing many lines of code. It is an open-source web-based interactive environment for H2O that allows you to combine code execution, text, mathematics, plots, and rich media in a single document. With H2O Flow, you can efficiently capture, rerun, annotate, present, and share your workflow[1]. In this chapter, we show you how to build and test a model using Python code. To install H2O in the Python environment use `pip install h2o` and to install it in the Conda environment use `conda install -c h2oai h2o`.

[1]`http://docs.h2o.ai/h2o/latest-stable/h2o-docs/flow.html`

© Tshepo Chris Nokeri 2021
T. C. Nokeri, *Data Science Revealed*, https://doi.org/10.1007/978-1-4842-6870-4_13

How H2O Works

H2O enables us to develop, test, and validate machine learning and deep learning models with little technical effort. When using this package, we do not write many lines of code. It automates most data science procedures. Listing 13-1 initializes H2O.

Listing 13-1. Initialize H2O

```
import h2o
h2o.init()
```

Table 13-1. *Environment Information*

H2O_cluster_uptime:	**07 secs**
H2O_cluster_timezone:	America/Los_Angeles
H2O_data_parsing_timezone:	UTC
H2O_cluster_version:	3.30.0.7
H2O_cluster_version_age:	3 months and 3 days
H2O_cluster_name:	H2O_from_python_i5_lenov_z4hv4c
H2O_cluster_total_nodes:	1
H2O_cluster_free_memory:	2.975 Gb
H2O_cluster_total_cores:	0
H2O_cluster_allowed_cores:	0
H2O_cluster_status:	accepting new members, healthy
H2O_connection_url:	`http://127.0.0.1:54321`
H2O_connection_proxy:	{"http": null, "https": null}
H2O_internal_security:	False
H2O_API_Extensions:	Amazon S3, Algos, AutoML, Core V3, TargetEncoder, Core V4
Python_version:	3.7.6 final

Data Processing

We load the data into an H2O dataframe using the `import_file()` method. The package supports different file formats. Likewise, we obtained the example data from Kaggle.[2]

Listing 13-2. Load Data into a H2O Dataframe

```
df = h2o.import_file(path_name)
```

Listing 13-3 returns first 10 data points.

Listing 13-3. View a H2O Dataframe

```
df.head()
```

Listing 13-4 assigns independent variables and the dependent variable to two separate dataframes.

Listing 13-4. Allocate Variables to X and Y Dataframe

```
y = "Outcome" x = df.col_names x.remove("Outcome") df["Outcome"] =
df["Outcome"].asfactor()
```

Listing 13-5 applies the `split_frame()` method to split data into training data, test data, and validation data.

Listing 13-5. Split Data into Training Data, Test Data and Validation Data

```
train, valid, test = df.split_frame(ratios=[.8, .1], seed=1234)
print("Train shape: ", train.shape)
print("Valid shape: ", valid.shape)
print("Test shape: ", test.shape)
```

Model Training

Listing 13-6 completes the logistic classifier.

[2]ihttps://www.kaggle.com/uciml/pima-indians-diabetes-database

Listing 13-6. Finalize the Logistic Classifier

```
from h2o.estimators.glm import H2OGeneralizedLinearEstimator
glm = H2OGeneralizedLinearEstimator(family="binomial")
glm.train(x=x, y=y, training_frame=train, validation_frame=valid)
```

We specified the family type as "binomial" since we are solving a binary classification problem. If we were solving a multiclass classification problem, then we would have specified the family type as "multinomial."

Model evaluation

Throughout the book, we depended on the confusion matrix, classification report, ROC curve, Precision-Recall curve, and learning curve to conclude about the binary classifiers' performance. The SciKit-Learn package provides a few matrices compared to the H2O package. Underneath, we adequately discuss classification evaluation metrics available in the H2O package, and we then show you how to estimate them. We begin with the Gini Index.

Gini Index

The H2O package enables us to obtain the Gini Index value, which measures the extent to which there is inequality in the values. We use the index to correctly determine the quality of a classifier. The index compromises values that range from 0 to 1. Where 0 shows that a classifier has perfect equality and 1 shows that a classifier has perfect inequality. When we develop a classifier, we ordinarily expect to find a Gini Index value that is closer to 1. Listing 13-7 returns estimates of the Gini index.

Listing 13-7. Gini Index

```
glm.gini(train=True, valid=True, xval=False)
{'train': 0.6828073360331426, 'valid': 0.5481481481481481}
```

The Gini Index value of the training data is 0.68 and the value of the validation data is 0.55. Both sets are on the borderline between equality and non-equality. We do not have perfect inequality, as expected. However, the values are not small enough to affect conclusions about the classifier.

Absolute Matthews Correlation Coefficient

Another classification model evaluation metric available on the H2O package is the Absolute Matthews Correlation Coefficient (MCC). We use it to determine the nature of the correlation between the actual classes and those predicted by a classifier. MCC has values that range from -1 to 1. Where -1 shows that a classifier makes many errors when predicting classes, 0 shows that a classifier struggles with guesstimating classes and 1 shows that a classifier is not skillful in distinguishing classes. Listing 13-8 returns estimates of the Absolute MCC.

Listing 13-8. Absolute MCC

```
glm.mcc(train=True, valid=True, xval=False)
{'train': [[0.3356601829569534, 0.5297584436640914]],
 'valid': [[0.3790626186045188, 0.39336862882432994]]}
```

The finding above shows that there is a weak positive correlation between actual classes and those predicted by the classifier.

Confusion Matrix

Listing 13-9 returns abstract information about a classifier's performance. (see Table 13-2).

Listing 13-9. Confusion Matrix

```
glm.confusion_matrix()
```

Table 13-2. *Confusion Matrix*

		0	1	Error	Rate
0	0	306.0	90.0	0.2273	(90.0/396.0)
1	1	49.0	168.0	0.2258	(49.0/217.0)
2	Total	355.0	258.0	0.2268	(139.0/613.0)

Listing 13-10 finds the F1 score of both the training data and validation data.

Listing 13-10. F1 Score Both the Training Data and Validation Data

```
glm.F1(train=True, valid=True, xval=False)
{'train': [[0.3356601829569534, 0.7073684210526315]],
 'valid': [[0.3790626186045188, 0.5405405405405405]]}
```

The findings above show that the logistic classifier has decent precession and recall. Listing 13-11 finds the accuracy coefficient.

Listing 13-11. Accuracy Coefficient

```
glm.accuracy(train=True, valid=True, xval=False)

{'train': [[0.5331122058852918, 0.7862969004893964]],
'valid': [[0.8926783510054471, 0.8115942028985508]]}
```

Listing 13-12 produces a curve that summarize how skillful the classifier is in distinguishing classes (see Figure 13-1).

Listing 13-12. ROC Curve

```
glm_perf.plot()
```

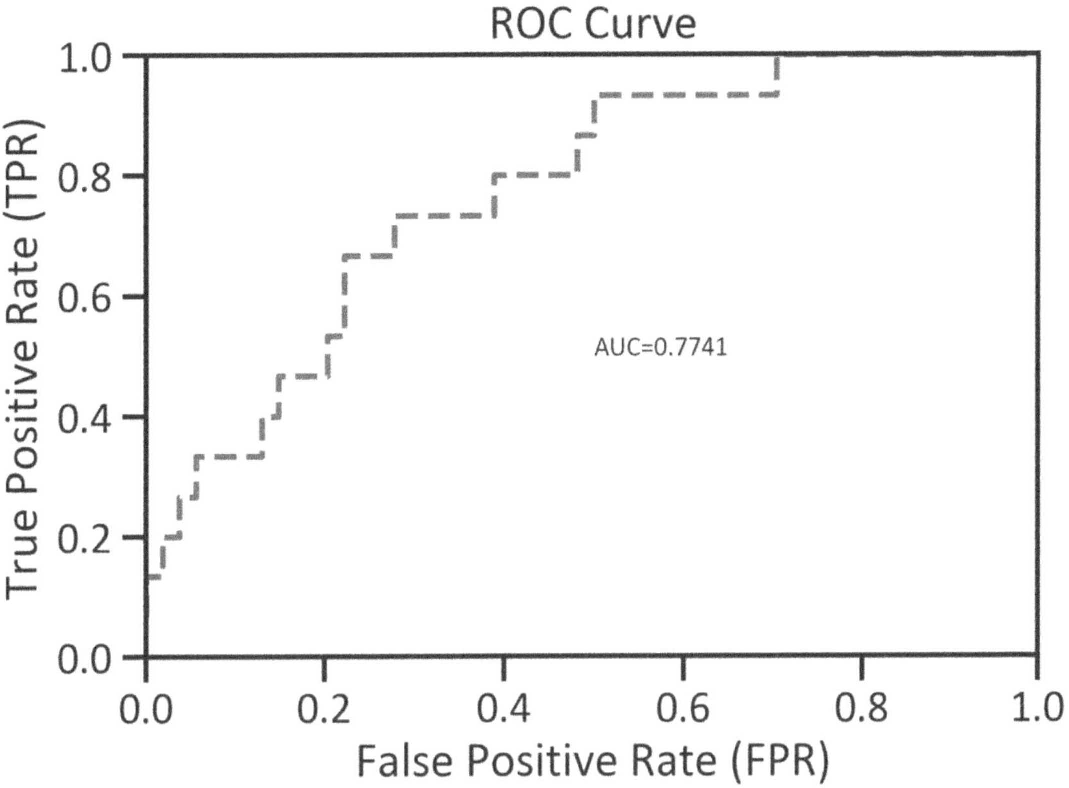

Figure 13-1. *ROC Curve*

Figure 13-1 exhibits the characteristics of a well-behaved curve; it follows the left-hand-side of the border in the beginning. However, as we increase the false positive rate, the true positive rate also increases.

Standardized Coefficient Magnitude

Listing 13-13 returns a plot that graphically represent the relationship between independent variables and the dependent variable (see Figure 13-2).

Listing 13-13. Standard Coefficient Magnitude

```
glm.std_coef_plot()
```

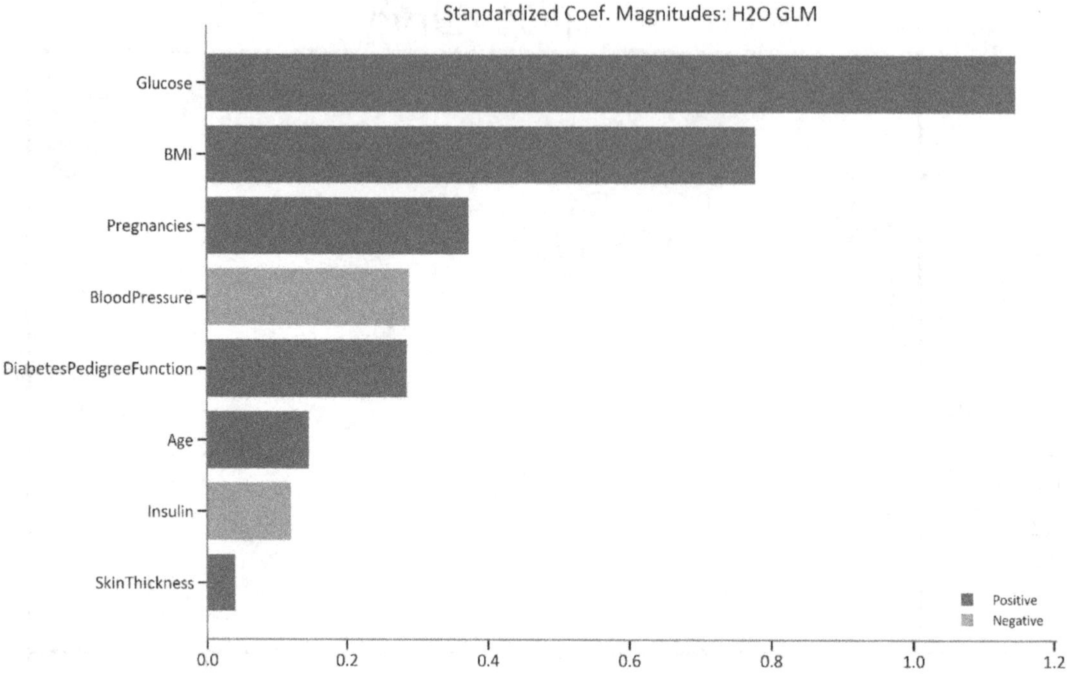

Figure 13-2. *Standardized Coefficient Magnitude*

Figure 13-2 shows the magnitude of the standardized coefficients. Most independent variables positively correlate with diabetes outcomes, besides blood pressure and insulin. We also see a strong positive correlation between glucose and diabetes outcomes.

Partial Dependence

We use partial dependence to determine the effect of the relationship between variables. Table 13-3 highlights the partial dependence between age and diabetes outcomes.

Listing 13-14. Partial Dependence

```
glm.partial_plot(data = df, cols = ["Age", "Outcome"], server=False,
plot = True)
```

Table 13-3. *Age Partial Dependence*

	Age	mean_response	stddev_response	std_error_mean_response
0	21.000000	0.322744	0.261743	0.009445
1	24.157895	0.328581	0.263071	0.009493
2	27.315789	0.334469	0.264344	0.009539
3	30.473684	0.340407	0.265559	0.009583
4	33.631579	0.346395	0.266716	0.009624

Table 13-3 and 13-4 highlight a change in values of the variables and the response mean, response standard deviation, and response standard error mean.

Table 13-4. *Glucose Partial Dependence*

	outcome	mean_response	stddev_response	std_error_mean_response
0	0	0.348501	0.272299	0.009826
1	1	0.348501	0.272299	0.009826

Figure 13-3 shows the variables and their corresponding mean response. It gives an idea of the partial dependence.

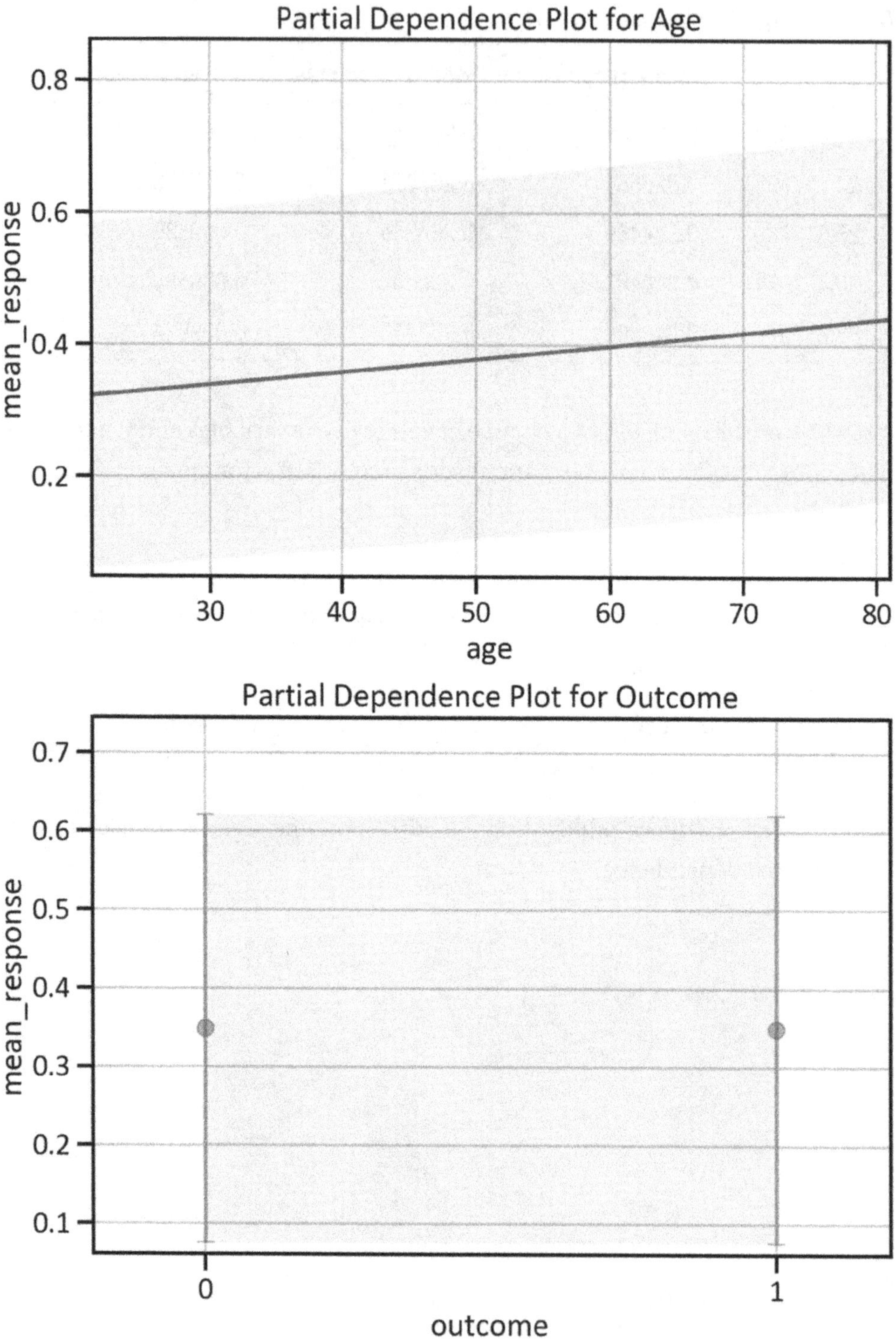

Figure 13-3. *Partial Dependence*

Feature Importance

We use feature importance as a dimension reduction technique by allocating each independent variable a value that shows the relative importance of each independent variable to the dependent variable. Listing 13-15 plots feature importance (see Figure 13-4).

Listing 13-15. Feature Importance

```
glm.varimp_plot()
```

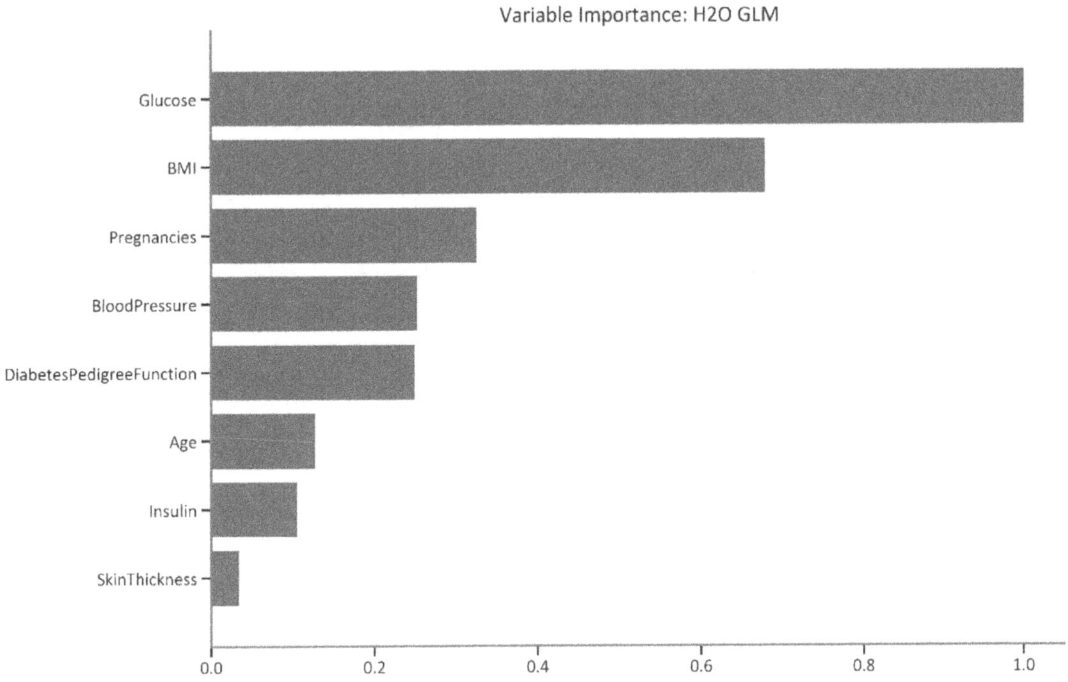

Figure 13-4. *Feature Importance*

Figure 13-4 shows that Glucose is the most important variable in diabetes outcomes. Followed by BMI and then pregnancies, and so forth. If we remove age, insulin, and Skin thickness, we can improve the performance of the classifier.

Predictions

Listing 13-16 tabulates the predicted classes and class probabilities (see Table 13-5)

Listing 13-16. Predicted Classes

```
y_predglm = glm.predict(test)
y_predglm
```

Table 13-5. *Predicted Classes And Class Probabilities*

predict	p0	p1
0	0.956033	0.0439669
0	0.977438	0.0225621
1	0.58329	0.41671
0	0.753834	0.246166
1	0.269132	0.730868
0	0.952432	0.0475679
0	0.811075	0.188925
0	0.917553	0.0824472
0	0.665598	0.334402
0	0.813001	0.186999

AutoML

AutoML stands for Automated Machine Learning. The term is self-explanatory. It enables us to build and rigorously test different machine learning and deep learning models using only a few line codes. It also automates processes like hyperparameter optimization and data modelling. Instead of developing many models and comparing them individually, we may use it. Developing an AutoML model is straightforward. Listing 13-17 completes the AutoML model.

Listing 13-17. Finalize the AutoML Model

```
from h2o.automl import H2OAutoML
automl = H2OAutoML(max_runtime_secs=120)
automl.train(x=x, y=y, training_frame=train, validation_frame=valid)
```

When we complete it, we must specify the maximum number of seconds that different models must take to learn the structure of the data. In our case, we specified 2 minutes. When we use it, in the background, the H2O package optimizes hyperparameters of available models and trains them.

Leaderboard

Table 13-6 highlights the performance of all models trained, including a 5-fold cross-validation model performance. It ranks the classifiers on the AUC score. A classifier first on the list has the highest AUC score, and a classifier last on the list has the lowest AUC score. It also highlights the log loss, the area under precision-recall, the mean value per class error, mean sum of errors and root mean sum of errors.

Listing 13-18. AutoML Leaderboard

```
leaderboard = automl.leaderboard
leaderboard
```

Table 13-6. *AutoML Leaderboard*

model_id	auc	logloss	aucpr	mean_per_class_error	rmse	mse
StackedEnsemble_BestOfFamily_AutoML_20201025_135923	0.830476	0.48388	0.703867	0.243867	0.397547	0.158043
GLM_1_AutoML_20201025_135923	0.829476	0.486136	0.70964	0.248761	0.397072	0.157666
GBM_grid__1_AutoML_20201025_135923_model_8	0.8278	0.483469	0.714668	0.244595	0.398872	0.159099
StackedEnsemble_AllModels_AutoML_20201025_135923	0.827573	0.486645	0.707761	0.248918	0.399255	0.159404
DeepLearning_grid__2_AutoML_20201025_135923_model_1	0.823174	0.506595	0.695583	0.244216	0.403943	0.16317
DeepLearning_grid__1_AutoML_20201025_135923_model_3	0.819415	0.571089	0.682411	0.257378	0.415348	0.172514
GBM_5_AutoML_20201025_135923	0.818548	0.494956	0.675015	0.244787	0.40579	0.164665
DeepLearning_grid__1_AutoML_20201025_135923_model_2	0.818118	0.554478	0.725002	0.237461	0.41781	0.174565
GBM_grid__1_AutoML_20201025_135923_model_9	0.815156	0.502735	0.687605	0.258513	0.408501	0.166873
GBM_grid__1_AutoML_20201025_135923_model_5	0.814243	0.498371	0.669688	0.262178	0.40745	0.166015

Table 13-6 shows that the leading binary classifier has an AUC score of 0.83 and an AUCPR of 0.70.

Prediction

Above, we found the leading classifier. Now let us consider the predicted classes and class probabilities (see Table 13-7).

Listing 13-19. Predicted Classes and Class Probabilities

```
y_pred = automl.predict(test)
y_pred
```

Table 13-7. *Predicted Classes and Class Probabilities*

predict	p0	p1
0	0.923024	0.0769757
0	0.896685	0.103315
1	0.625636	0.374364
1	0.713157	0.286843
1	0.37774	0.62226
0	0.922715	0.0772854
0	0.836983	0.163017
0	0.910994	0.0890061
1	0.587552	0.412448
1	0.686241	0.313759

Conclusion

This chapter introduced a driverless open-source package known as H2O. We developed and tested a logistic classifier and AutoML model. You might have noticed how intuitive and rich this package is. It does not require one to have extensive technical skills and experience. The ease of the package enables organizations to escalate adoption of practical AI solutions. We expect to see more driverless open-source packages in the foreseeable future.

We have now come to the end of the chapter that concludes the book; we believe you have gained sufficient insight into the world of data science. Now, apply the knowledge gained to solve complex problems. Always keep at abreast of the latest research and technologies. Never fall into the trap of deploying unreliable models or abandoning the scalability of models and web apps.

Index

© Tshepo Chris Nokeri 2021
T. C. Nokeri, *Data Science Revealed*, https://doi.org/10.1007/978-1-4842-6870-4

D

E

F

G

H

M

N, O

P, Q

R

S

CPSIA information can be obtained
at www.ICGtesting.com
Printed in the USA
LVHW101143110421
684149LV00006B/196